Mark Anthony Benvenuto
Industrial Chemistry

T0092553

Also of Interest

Industrial Chemistry.
For Advanced Students
Benvenuto, 2023
ISBN 978-3-11-077874-8, e-ISBN 978-3-11-077876-2

Materials Chemistry.
For Scientists and Engineers
Benvenuto, 2022
ISBN 978-3-11-065673-2, e-ISBN 978-3-11-065677-0

Chemistry and Energy.
From Conventional to Renewable
Benvenuto, 2022
ISBN 978-3-11-066226-9, e-ISBN 978-3-11-066227-6

Process Technology.
An Introduction
De Haan, Padding, 2022
ISBN 978-3-11-071243-8, e-ISBN 978-3-11-071244-5

Chemical Reaction Technology
Murzin, 2022
ISBN 978-3-11-071252-0, e-ISBN 978-3-11-071255-1

Mark Anthony Benvenuto

Industrial Chemistry

—

2nd edition

DE GRUYTER

Author
Prof. Dr. Mark Anthony Benvenuto
University of Detroit Mercy
Department of Chemistry & Biochemistry
4001 W. McNichols Rd.
Detroit MI 48221-3038
United States of America
benvenma@udmercy.edu

ISBN 978-3-11-067106-3
e-ISBN (PDF) 978-3-11-067109-4
e-ISBN (EPUB) 978-3-11-067121-6

Library of Congress Control Number: 2023932712

Bibliographic information published by the Deutsche Nationalbibliothek
The Deutsche Nationalbibliothek lists this publication in the Deutsche Nationalbibliografie;
detailed bibliographic data are available on the Internet at http://dnb.dnb.de.

© 2023 Walter de Gruyter GmbH, Berlin/Boston
Cover image: Alecio Cezar / iStock / Getty Images Plus
Typesetting: VTeX UAB, Lithuania
Printing and binding: CPI books GmbH, Leck

www.degruyter.com

Preface

It is hoped that a book like this will find a place in the chemistry and engineering curriculum simply because there do not seem to be many books out there that look at the huge world of industrial scale chemistry. And yet, this is a truly fascinating part of our world. It enables so much. It helps feed 7 billion people. It provides us with medicines. It gives us clothing, household end products, and transportation possibilities that have never existed before. It is not exaggeration to say that the Industrial Revolution, and the part that chemical processes have played in it, moved the people of our planet from an agrarian existence to that which we enjoy today.

So, why is all this important? Because the processes and reactions discussed here enable modern life, and the quality of life people now enjoy.

Interestingly, quite a few historians have looked at events and developments that have shaped the transition from the ancient to the modern world. Numerous books have been written about how a catastrophe changed the world, or the onset of a plague, or the death of a single monarch at a specific point in time. Few of these authors are also chemists who choose to look at how the chemical processes we have harnessed have changed our world.

As well, a thorough examination of what we do in terms of large scale chemistry can provide us with ideas on how to improve. Each chapter of this book looks not only at a series of chemical processes, but also at the pollution that can be generated, the possibilities for recycling starting and end-use materials, and thus at possible ways to phase out pollutants and re-use materials.

This book then arises from a perceived need for an understandable text that might be used in an undergraduate or graduate setting for a course on industrial chemistry. It can also serve as a reference for any curriculum that meshes chemistry and engineering.

The biggest difficulty in producing a volume such as this is determining what to include, and what to leave out. We have ensured that the large process chemistry which enables so much of our modern way of life is included here: the production of sulfuric acid, ammonia, the derivatives of limestone, and of course the products made from petroleum feed stocks. But there are others as well, and there are some topics that we chose to omit. The decision to leave something out is always difficult, but hopefully, the book will prove itself useful.

Finally, there are always a large number of people to thank when such an endeavor as this is completed. My editors, Julia Lauterbach and Karin Sora have been incredibly helpful at every step of the way. My work colleagues, Drs. Matt Mio, Kendra Evans, Klaus Friedrich, Kate Lanigan, Liz Roberts-Kirchhoff, Jon Stevens, Mary Lou Caspers, and Shula Schlick, as well as Jane Schley and Meghann Murray have all been valuable sources of information, and have all had a hand at answering the multitude of questions I generated while writing this. Friends and colleagues at companies such as BASF, Ash Stevens and GM, including Heinz Plaumann, Hulya Ahmed, Jay Otten, Megan Klein, Kevin Perry, and Keith Olsen have also been very kind in putting up with my endless

https://doi.org/10.1515/9783110671094-201

questions over the past months. And of course, my wife Marye and my sons David and Christian come quickest to mind when I think of the people who had to put up with all the problems, complaining, and late night work and writing sessions any author choses to do. Thank you all.

Detroit   Mark A. Benvenuto

Contents

1 Overview and introduction to the chemical industry

1.1 Introduction

There have been numerous advances in the quality of life for humankind over the course of the centuries, but one can argue that the emergence of a chemical industry in the early to mid-nineteenth century, as part of the growing Industrial Revolution, has helped in countless ways. While the production of sulfuric acid, for example, may seem removed from direct improvements to human health and well-being, it is used in processes that result in a wide variety of the consumer products we have available to us. The fertilizer used to grow crops, the metals that are extracted and refined from the earth, and many other processes involve starting chemicals such as sulfuric acid. All of these in turn result in products and materials that make human life better today than in the time before a chemical industry.

We will examine the chemical industry from the point of view of the chemist, with an emphasis on reaction chemistry, as opposed to that of the businessman or other professionals. We will examine the chemical reactions and operations that produce the materials used on an industrial scale.

1.2 Listing of U. S. top 50 chemical producers

An examination of the top 50 U. S. chemical producers, ranked annually, shows numerous ones that do not appear to be makers of direct household or consumer materials. This is because roughly one quarter of chemical production is used by other chemical companies for the production of still other products. In addition, numerous small companies and niche or specialty companies do not make the list because the volume or dollar sales of their product(s) are not high enough.

A look at these companies reveals that many of them are involved in the direct production of a few basic commodity chemicals, or the production of components that then go into making such chemicals, or the consumer end use products involving some transformation of these basic chemicals. The materials in question include: polyethylene of various densities, polypropylene, polyvinyl chloride, and polystyrene, as well as polyethylene terephthalate. These five materials account for more than $\frac{1}{2}$ of bulk chemical production worldwide.

Figures 1.1 and 1.2 show the Top 50 US chemicals producers for the previous years.

It should be noted that some extremely large firms that have a major presence within the US, such as BASF, Air Liquide, and Akzo Nobel, are not listed in Figures 1.1 and 1.2 because their major headquarters are not in the United States. These are shown in Figure 1.3.

https://doi.org/10.1515/9783110671094-001

Global Top 50

Chemical sales and profits swelled for nearly every company in 2021.

RANK 2021	RANK 2020[a]	COMPANY	CHEMICAL SALES ($ MILLIONS)	CHANGE FROM 2020	CHEMICAL SALES AS % OF TOTAL SALES	SECTOR	CHEMICAL OPERATING PROFIT[b] ($ MILLIONS)	CHANGE FROM 2020	CHEMICAL OPERATING PROFIT MARGIN[c]	HEADQUARTERS	IDENTIFIABLE CHEMICAL ASSETS ($ MILLIONS)	OPERATING RETURN ON CHEMICAL ASSETS[d]
1	1	BASF	$92,982	32.9%	100.0%	Diversified	$9,179	80.5%	9.9%	Germany	$103,375	8.9%
2	2	Sinopec	65,848	31.9	15.9	Petrochemicals	1,761	9.5	2.7	China	34,539	5.1
3	3	Dow	54,968	42.5	100.0	Diversified	7,887	208.6	14.3	US	62,990	12.5
4	5	Sabic	43,230	50.1	92.7	Petrochemicals	8,779	445.8	20.3	Saudi Arabia	79,919	11.0
5	4	Formosa Plastics[e]	43,173	47.8	72.2	Petrochemicals	n/a	n/a	n/a	Taiwan	n/a	n/a
6	16	Ineos	39,937	121.2	100.0	Diversified	5,370	344.2	13.4	UK	37,226	14.4
7	6	PetroChina	39,593	41.7	9.8	Petrochemicals	1,862	9.5	4.7	China	n/a	n/a
8	11	LyondellBasell Industries	38,995	66.6	84.5	Petrochemicals	8,009	172.6	20.5	US	n/a	n/a
9	7	LG Chem	37,257	41.8	100.0	Diversified	4,389	179.5	11.8	South Korea	44,664	9.8
10	12	ExxonMobil	36,856	59.6	13.3	Petrochemicals	9,960	272.3	27.0	US	39,722	25.1
11	8	Mitsubishi Chemical Group	30,719	24.8	84.8	Diversified	2,547	74.3	8.3	Japan	43,120	5.9
12	13	Hengli Petrochemical[f]	27,961	31.9	91.1	Petrochemicals	n/a	n/a	n/a	China	n/a	n/a
13	9	Linde	27,926	14.5	90.7	Industrial gases	6,703	25.0	24.0	UK	n/a	n/a
14	10	Air Liquide	27,148	13.4	98.3	Industrial gases	2,779	16.3	10.2	France	50,645	5.5
15	14	Syngenta Group	24,900	20.9	81.1	Agricultural chemicals	n/a	n/a	n/a	Switzerland	n/a	n/a
16	20	Reliance Industries[e]	22,583	65.6	21.1	Petrochemicals	n/a	n/a	n/a	India	n/a	n/a
17	27	Wanhua Chemical	22,561	98.2	100.0	Diversified	4,978	142.1	22.1	China	29,502	16.9
18	29	Braskem	19,575	80.4	100.0	Petrochemicals	5,038	278.0	25.7	Brazil	17,155	29.4
19	17	Sumitomo Chemical	19,176	24.7	76.2	Diversified	1,581	118.3	8.2	Japan	23,747	6.7
20	21	Shin-Etsu Chemical[f]	18,885	38.6	100.0	Diversified	6,157	72.4	32.6	Japan	36,901	16.7
21	22	Covestro	18,813	48.5	100.0	Diversified	2,655	206.6	14.1	Germany	18,421	14.4
22	18	Toray Industries	17,856	20.9	88.0	Diversified	1,227	40.3	6.9	Japan	n/a	n/a
23	19	Evonik Industries	17,692	22.6	100.0	Specialty chemicals	1,541	39.7	8.7	Germany	26,362	5.8
24	23	Shell	16,993	45.0	6.5	Petrochemicals	1,390	72.0	8.2	UK	n/a	n/a
25	15	DuPont	16,653	-18.4	100.0	Specialty chemicals	2,652	59.7	15.9	US	45,707	5.8
26	25	Yara	16,517	43.4	100.0	Fertilizers	1,068	-9.2	6.4	Norway	17,272	6.2
27	34	Rongsheng Petrochemical	16,001	59.6	58.3	Petrochemicals	n/a	n/a	n/a	China	n/a	n/a
28	31	Lotte Chemical	15,827	48.2	100.0	Diversified	1,341	330.3	8.5	South Korea	19,976	6.7
29	28	Mitsui Chemicals	14,661	33.1	100.0	Diversified	1,269	70.3	8.6	Japan	17,615	7.2
30	32	Indorama Ventures	14,626	41.2	100.0	Petrochemicals	1,315	339.6	9.0	Thailand	16,929	7.8
31	42	Chevron Phillips Chemical	14,104	67.1	100.0	Petrochemicals	n/a	n/a	n/a	US	17,777	n/a
32	33	Umicore	13,567	34.4	47.7	Specialty chemicals	542	104.3	4.0	Belgium	9,134	5.9
33	26	Solvay	13,527	17.7	100.0	Specialty chemicals	1,618	37.5	12.0	Belgium	23,718	6.8
34	24	Bayer	12,743	9.7	24.4	Agricultural chemicals	n/a	n/a	n/a	Germany	n/a	n/a
35	40	Mosaic	12,357	42.3	100.0	Fertilizers	2,770	299.5	22.4	US	22,035	12.6
36	46	Nutrien	11,590	62.0	41.8	Fertilizers	4,825	242.9	41.6	Canada	25,940	18.6
37	36	Arkema	11,261	20.7	100.0	Specialty chemicals	1,320	98.6	11.7	France	14,552	9.1
38	37	Asahi Kasei	10,908	20.9	48.7	Specialty chemicals	1,004	65.9	9.2	Japan	16,214	6.2
39	35	DSM	10,888	13.5	100.0	Specialty chemicals	1,247	38.7	11.5	Netherlands	18,944	6.6
39	38	Hanwha Solutions[f]	10,888	22.8	86.6	Specialty chemicals	541	12.3	5.0	South Korea	17,465	3.1
41	41	Eastman Chemical	10,476	23.6	100.0	Specialty chemicals	1,451	32.5	13.9	US	15,519	9.3
42	30	Johnson Matthey	10,412	-2.5	47.2	Specialty chemicals	387	-9.1	3.7	UK	3,204	12.1
43	39	Air Products	10,323	16.6	100.0	Industrial gases	2,215	3.6	21.5	US	26,859	8.2
44	—	EuroChem Group	10,202	65.8	100.0	Fertilizers	3,400	170.9	33.3	Switzerland	14,269	23.8
45	44	Borealis	10,164	26.0	100.0	Petrochemicals	1,668	435.9	16.4	Austria	15,361	10.9
46	—	PTT Global Chemical	9,084	52.1	62.0	Diversified	1,140	483.3	12.5	Thailand	16,521	6.9
47	43	Sasol	9,011	10.8	65.9	Diversified	1,337	n.m.	14.8	South Africa	n/a	n/a
48	—	TongKun Group[f]	8,996	28.4	100.0	Petrochemicals	n/a	n/a	n/a	China	n/a	n/a
49	45	Lanxess	8,940	23.8	100.0	Specialty chemicals	562	16.4	6.3	Germany	12,443	4.5
50	—	Hengyi Petrochemical	8,856	66.8	44.3	Diversified	n/a	n/a	n/a	China	n/a	n/a

Sources: Company documents, C&EN analysis. **Note:** Some figures converted at 2021 average exchange rates of US$1.00= 5.3958 Brazilian reais, 6.4508 Chinese yuan, 0.8453 euros, 73.9351 Indian rupees, 109.8429 Japanese yen, 1144.8911 South Korean won, 3.75 Saudi riyals, 14.7751 South African rand, 27.9366 New Taiwan dollars, and 32.0052 Thai baht. *n/a* means not available, and *n.m.* means not meaningful. **a** Prior-year rankings have been revised from the July 26, 2021, issue of C&EN to reflect restated results and changes in exchange rates. **b** Chemical sales less administrative expenses and cost of sales. **c** Chemical operating profit as a percentage of chemical sales. **d** Chemical operating profit as a percentage of identifiable chemical assets. **e** C&EN estimates. **f** Chemical sales include a significant amount of nonchemical products.

Figure 1.1: Top 50 Chemical Producers for 2021 (courtesy of Chemical & Engineering News) [6].

1.3 List of top chemical producers worldwide

Once again, the top chemical companies shown in Figure 1.3 may not be household names, but they are all familiar to chemists and chemical engineers.

Several of these firms may switch spots on any list over the course of years, but as with US firms, changes tend to be minimal. BASF (more completely, Badische Anilin und

TOP 50 U.S. CHEMICAL FIRMS
Most firms experienced strong increases in sales and profits in 2011

RANK 2011	2010	COMPANY	CHEMICAL SALES ($ MILLIONS) 2011	CHANGE FROM 2010	CHEMICAL SALES AS % OF TOTAL SALES	HEADQUARTERS	CHEMICAL OPERATING PROFITS[a] ($ MILLIONS)	CHANGE FROM 2010	CHEMICAL OPERATING PROFITS AS % OF TOTAL OPERATING PROFIT	OPERATING PROFIT MARGIN[b]	IDENTIFIABLE CHEMICAL ASSETS ($ MILLIONS)	CHEMICAL ASSETS AS % OF TOTAL ASSETS	CHEMICAL OPERATING RETURN ON CHEMICAL ASSETS[c]
1	1	Dow Chemical	$59,985	11.8%	100.0%	Midland, Mich.	$4,522	24.7%	100.0%	7.5%	$69,224	100.0%	6.5%
2	2	ExxonMobil[d]	41,942	18.1	9.0	Irving, Texas	4,383	-10.8	10.7	10.5	27,107	8.2	16.2
3	3	DuPont[e]	34,763	15.6	91.6	Wilmington, Del.	5,547	22.8	98.8	16.0	18,819	67.9	29.5
4	5	Chevron Phillips	13,935	24.4	100.0	The Woodlands, Texas	na	na	na	na	8,634	100.0	na
5	4	PPG Industries	13,824	11.1	92.9	Pittsburgh	1,930	17.1	95.2	14.0	10,557	73.4	18.3
6	6	Praxair[e]	11,252	11.2	100.0	Danbury, Conn.	3,465	12.2	100.0	30.8	16,356	100.0	21.2
7	7	Huntsman Corp.	11,221	21.3	100.0	Salt Lake City	753	67.7	100.0	6.7	8,657	100.0	8.7
8	10	Mosaic[f]	9,938	47.0	100.0	Plymouth, Minn.	2,749	106.2	100.0	27.7	15,787	100.0	17.4
9	8	Air Products[g]	9,681	13.2	96.0	Allentown, Pa.	1,624	13.9	96.3	16.8	13,035	97.3	12.5
10	9	Momentive	7,844	5.9	100.0	Columbus, Ohio	771	-12.6	100.0	9.8	6,273	100.0	12.3
11	13	Eastman Chemical	7,178	22.9	100.0	Kingsport, Tenn.	1,013	13.7	100.0	14.1	6,184	100.0	16.4
12	12	Celanese	6,763	14.3	100.0	Dallas	802	32.6	100.0	11.9	8,518	100.0	9.4
13	11	Dow Corning	6,427	7.2	100.0	Midland, Mich.	na	na	na	na	13,571	100.0	na
14	14	Lubrizol	6,100	12.6	100.0	Wickliffe, Ohio	na	na	na	na	na	na	na
15	20	CF Industries	6,098	53.8	100.0	Long Grove, Ill.	2,766	157.6	100.0	45.4	8,975	100.0	30.8
16	15	Styron	6,000	17.7	100.0	Berwyn, Pa.	na	na	na	na	na	na	na
17	16	Honeywell[e]	5,659	19.7	15.5	Morristown, N.J.	1,042	39.1	19.5	18.4	5,402	13.6	19.3
18	18	Occidental Petroleum	4,815	19.9	20.1	Los Angeles	861	96.6	7.9	17.9	3,754	6.3	22.9
19	17	Ecolab	4,649	9.4	41.2	St. Paul	na	na	na	na	na	na	na
20	19	Ashland[g]	4,531	13.7	69.7	Covington, Ky.	301	19.4	58.6	6.6	9,792	75.5	3.1
21	21	Westlake Chemical	3,620	14.1	100.0	Houston	447	18.1	100.0	12.3	3,267	100.0	13.7
22	22	FMC Corp.	3,378	8.4	100.0	Philadelphia	620	15.0	100.0	18.3	3,744	100.0	16.6
23	25	Monsanto[h]	3,240	12.1	27.4	St. Louis	281	nm	11.8	8.7	4,493	22.6	6.3
24	28	W.R. Grace	3,212	20.1	100.0	Columbia, Md.	525	40.4	100.0	16.3	4,497	100.0	11.7
25	24	Cabot Corp.[g]	3,102	7.2	100.0	Boston	243	-2.0	100.0	7.8	3,141	100.0	7.7
26	27	Cytec Industries	3,073	11.8	100.0	Woodland Park, N.J.	308	13.6	100.0	10.0	3,537	100.0	8.7
27	29	Rockwood Specialties	3,053	15.3	83.2	Princeton, N.J.	451	90.5	81.5	14.8	3,910	85.2	11.5
28	26	Chemtura	3,025	9.6	100.0	Philadelphia	347	15.7	100.0	11.5	2,855	100.0	12.2
29	30	Albemarle	2,869	21.4	100.0	Richmond, Va.	588	39.4	100.0	20.5	3,204	100.0	18.3
30	35	TPC Group	2,759	43.8	100.0	Houston	128	24.1	100.0	4.7	970	100.0	13.2
31	31	Georgia Gulf	2,575	13.4	79.9	Atlanta	154	11.7	95.3	6.0	996	60.6	15.4
32	32	Ferro Corp.[e]	2,156	2.6	100.0	Mayfield Heights, Ohio	118	-28.2	100.0	5.5	1,441	100.0	8.2
33	34	NewMarket	2,138	19.7	100.0	Richmond, Va.	302	5.1	100.0	14.1	1,192	100.0	25.3
34	33	Solutia	2,097	7.5	100.0	St. Louis	360	9.4	100.0	17.2	3,526	100.0	10.2
35	37	Kronos Worldwide	1,943	34.0	100.0	Dallas	553	209.9	100.0	28.5	1,824	100.0	30.3
36	38	Stepan	1,843	28.8	100.0	Northfield, Ill.	119	9.8	100.0	6.4	901	100.0	13.2
37	50	Tronox	1,651	35.7	100.0	Oklahoma City	307	89.2	100.0	18.6	1,657	100.0	18.5
38	43	Goodyear	1,594	41.0	7.0	Akron, Ohio	na	na	na	na	na	na	na
39	41	H.B. Fuller[e]	1,558	14.9	100.0	St. Paul	129	23.7	100.0	8.3	1,228	100.0	10.5
40	40	Sigma-Aldrich	1,503	10.3	60.0	St. Louis	na	na	na	na	na	na	na
41	42	Kraton Polymers	1,438	17.0	100.0	Houston	187	1.1	100.0	13.0	1,154	100.0	16.2
42	46	Olin	1,389	34.0	70.8	Clayton, Mo.	245	109.0	64.6	17.6	1,652	67.5	14.8
43	47	Reichhold	1,183	14.6	100.0	Research Triangle Park, N.C.	na	na	na	na	na	na	na
44	45	OM Group	1,111	2.5	73.4	Cleveland	143	-7.6	161.9	12.9	1,344	46.8	10.7
45	44	Sunoco	1,095	-21.9	2.3	Philadelphia	1	-98.2	0.8	0.1	na	na	na
46	48	Koppers	1,016	27.7	66.0	Pittsburgh	45	-41.5	56.6	4.5	495	67.8	9.2
47	—	Omnova[i]	952	80.3	79.3	Fairlawn, Ohio	87	18.0	101.5	9.1	564	65.1	15.4
48	49	Innophos	811	13.5	100.0	Cranbury, N.J.	137	43.6	100.0	16.9	687	100.0	19.9
48	—	MeadWestvaco	811	19.4	13.4	Richmond, Va.	203	44.0	24.2	25.0	460	5.2	44.1
50	—	Innospec	774	13.3	100.0	Littleton, Colo.	52	-37.1	100.0	6.7	569	100.0	9.2

a Operating profit is sales less administrative expenses and cost of sales. b Operating profit as a percentage of sales. c Chemical operating profit as a percentage of identifiable assets. d Profits and profitability ratios are after tax. e Sales include a significant amount of nonchemical products. f Fiscal year ended May 31. g Fiscal year ended Sept. 30. h Fiscal year ended Aug. 31. i Fiscal year ended Dec. 3. j Fiscal year ended Nov. 30. na – not available. nm – not meaningful.

Figure 1.2: Top 50 Chemical Producers for 2011 (courtesy of Chemical & Engineering News).

Soda Fabrik) has been the single largest producer now for several years. Some firms do not make the list because their overall sales are not exclusively in chemicals. Alphabetically, a list of top producers worldwide can include [2, 3, 4, 5, 1, 7, 8, 9, 10, 11, 12, 13, 14, 15, 16, 17]:
- Arkema
- BASF
- Bayer

Top 50 US chemical firms

Coming out of pandemic-depressed 2020, chemical companies saw ginormous increases in sales and profits.

RANK 2021	RANK 2020a	COMPANY	CHEMICAL SALES ($ MILLIONS)	CHANGE FROM 2020	CHEMICAL SALES AS % OF TOTAL SALES	SECTOR	CHEMICAL OPERATING PROFIT b ($ MILLIONS)	CHANGE FROM 2020	OPERATING PROFIT MARGIN c	HEADQUARTERS	IDENTIFIABLE CHEMICAL ASSETS ($ MILLIONS)	OPERATING RETURN ON CHEMICAL ASSETS d
1	1	Dow	$54,968	42.6%	100.0%	Diversified	$7,887	208.6%	14.3%	Midland, Michigan	$62,990	12.5%
2	2	LyondellBasell Industries	38,995	66.6	84.5	Petrochemicals	8,009	172.6	20.5	Houston	n/a	n/a
3	3	ExxonMobil	36,858	59.6	13.3	Petrochemicals	9,960	272.3	27.0	Irving, Texas	39,722	25.1
4	4	DuPont	16,653	-18.4	100.0	Diversified	2,652	59.7	15.9	Wilmington, Delaware	45,707	5.8
5	8	Chevron Phillips Chemical	14,104	67.1	100.0	Petrochemicals	n/a	n/a	n/a	The Woodlands, Texas	17,777	n/a
6	6	Mosaic	12,357	42.3	100.0	Fertilizers	2,770	299.5	22.4	Tampa, Florida	22,036	12.6
7	7	Eastman Chemical	10,476	23.6	100.0	Diversified	1,451	32.5	13.9	Kingsport, Tennessee	15,519	9.3
8	5	Air Products e	10,323	16.6	100.0	Industrial gases	2,215	3.6	21.5	Allentown, Pennsylvania	26,859	8.2
9	14	Westlake	8,670	58.6	73.5	Petrochemicals	2,549	1,003.5	29.4	Houston	11,938	21.4
10	13	Celanese	8,537	51.0	100.0	Diversified	1,936	171.0	22.7	Irving, Texas	11,975	16.2
11	12	Huntsman	8,256	39.9	100.0	Diversified	774	307.4	9.4	The Woodlands, Texas	9,392	8.2
12	17	Olin	7,327	51.7	82.2	Chlorine chemistry	1,614	3,544.0	22.0	Clayton, Missouri	7,492	21.5
13	9	Corteva Agriscience	7,253	12.3	46.3	Agrochemicals	825	32.0	11.4	Indianapolis, Indiana	12,428	6.6
14	19	CF Industries	6,538	58.5	100.0	Fertilizers	2,164	263.7	33.1	Deerfield, Illinois	12,375	17.5
15	11	Lubrizol	6,500	9.2	100.0	Specialties	490	-51.0	7.5	Wickliffe, Ohio	n/a	n/a
16	15	Chemours	6,345	27.7	100.0	Diversified	682	52.6	10.7	Wilmington, Delaware	7,550	9.0
17	10	Ecolab f	6,305	5.8	49.5	Process services	1,031	-6.8	16.4	Saint Paul, Minnesota	n/a	n/a
18	16	Honeywell International	5,402	11.8	15.7	Fluorochemicals	n/a	n/a	n/a	Charlotte, North Carolina	n/a	n/a
19	20	Occidental Petroleum	5,245	40.5	20.2	Petrochemicals	1,544	132.5	29.4	Houston	4,671	33.1
20	18	FMC	5,045	8.7	100.0	Agrochemicals	1,153	11.5	22.9	Philadelphia	10,561	10.9
21	22	Trinseo	4,828	59.0	100.0	Polymers	376	494.1	7.8	Berwyn, Pennsylvania	4,712	8.0
22	24	Tronox	3,572	29.5	100.0	Pigments	577	110.6	16.2	Stamford, Connecticut	5,967	9.6
23	25	Cabot e	3,409	30.4	100.0	Specialties	454	200.7	13.3	Boston	3,306	13.7
24	21	Albemarle	3,328	6.4	100.0	Specialties	503	-0.7	15.1	Charlotte, North Carolina	10,974	4.6
25	23	H.B. Fuller fg	3,278	17.5	100.0	Specialties	253	15.7	7.7	Saint Paul, Minnesota	4,275	5.9
26	26	Hexion h	2,550	1.5	100.0	Specialties	237	426.7	9.3	Columbus, Ohio	3,764	6.3
27	28	Avantor	2,548	24.7	34.5	Laboratory chemicals	n/a	n/a	n/a	Radnor, Pennsylvania	n/a	n/a
28	34	Avient i	2,392	59.8	49.6	Pigments	303	67.6	12.7	Avon Lake, Ohio	2,965	10.2
29	29	NewMarket	2,356	17.2	100.0	Fuel additives	256	-17.3	10.9	Richmond, Virginia	2,558	10.1
30	30	Stepan	2,346	25.5	100.0	Detergents	161	-0.9	7.7	Northbrook, Illinois	2,066	8.8
31	—	International Flavors & Fragrances	2,329	1,638.1	20.0	Food additives	272	6,700.0	11.7	New York City	14,774	1.8
32	27	Ashland e	2,111	-9.2	100.0	Specialties	172	28.4	8.1	Wilmington, Delaware	6,612	2.6
33	42	ChampionX f	1,984	88.1	64.5	Oil field chemicals	196	122.2	9.9	The Woodlands, Texas	n/a	n/a
34	33	Kraton	1,970	26.0	100.0	Polymers, pine chemicals	258	272.4	13.1	Houston	2,655	9.7
35	32	Kronos Worldwide	1,939	18.3	100.0	Pigments	197	48.8	10.2	Dallas	2,013	9.8
36	40	Americas Styrenics	1,822	63.3	100.0	Polymers	205	135.6	11.3	The Woodlands, Texas	702	29.2
37	37	AdvanSix	1,665	45.5	100.0	Polymers	191	203.8	11.3	Parsippany, New Jersey	1,312	14.6
38	38	Orion Engineered Carbons	1,547	36.1	100.0	Inorganics	154	60.6	10.0	Houston	1,631	9.5
39	36	Innospec	1,483	24.3	100.0	Fuel additives	130	74.2	8.8	Englewood, Colorado	1,571	8.3
40	35	Ingevity	1,392	14.4	100.0	Pine chemicals	307	4.7	22.1	North Charleston, South Carolina	2,469	12.4
41	43	3M	1,205	16.3	3.4	Fluorochemicals	n/a	n/a	n/a	Saint Paul, Minnesota	n/a	n/a
42	39	CMC Materials e	1,200	7.5	100.0	Electronic materials	215	-1.9	17.9	Aurora, Illinois	2,151	10.0
43	44	Ferro	1,126	17.4	100.0	Pigments	127	38.9	11.3	Mayfield Heights, Ohio	1,290	9.8
44	46	Genesis Energy	973	9.9	45.6	Inorganics	167	28.2	17.1	Houston	2,133	7.8
45	45	Koppers	949	4.3	56.5	Coal tar chemicals	143	27.8	15.0	Pittsburgh	1,034	13.8
46	—	Entegris i	711	16.7	30.9	Electronic materials	166	31.1	23.6	Billerica, Massachusetts	1,192	14.1
47	47	Balchem	670	13.0	83.8	Nutritional ingredients	102	11.8	15.3	New Hampton, New York	886	11.5
48	41	Ecovyst	611	-44.6	100.0	Inorganics	79	-46.7	12.9	Malvern, Pennsylvania	1,931	4.1
49	49	Minerals Technologies	579	13.3	31.2	Inorganics	73	7.5	12.6	New York City	606	12.0
50	—	Goodyear Tire & Rubber	569	79.5	5.7	Polymers	n/a	n/a	n/a	Akron, Ohio	n/a	n/a

Sources: Company documents, C&EN analysis. **Note:** n/a means not available. **a** Prior-year ranking has been restated to reflect the inclusion of only chemical producing segments at Westlake following nonchemical acquisitions. **b** Chemical sales minus administrative expenses and cost of sales. **c** Chemical operating profit as a percentage of chemical sales. **d** Chemical operating profit as a percentage of identifiable chemical assets. **e** Fiscal year ended Sept. 30, 2021. **f** Chemical sales include a significant amount from nonchemical products. **g** Fiscal year ended Nov. 27, 2021. **h** Figures are for the first 9 months because the company was acquired in early 2022 and didn't publish an annual report.

Figure 1.3: Top Chemical Companies Worldwide 2021 (courtesy of Chemical & Engineering News).

- Braskem
- Celanese
- Degussa
- Dow
- Dupont
- Eastman Chemical

- INEOS
- LyondellBasell
- Mitsubishi
- Pittsburgh Plate Glass (PPG)
- Saudi Basic Chemicals (SABIC)
- Shell
- Wanhua

1.4 Employment

The chemical industry employs literally millions of people, from factory and process employees who ensure that bulk, basic chemicals are manufactured properly, to inspectors, quality control, and quality assurance personnel, to PhD chemists and chemical engineers who lead research teams within the research and development (R&D) arms of companies. Different firms require different numbers of various personnel, not simply because of the company size, but because of the output and products of each firm. For example, the sulfuric acid industry is very mature, and does not require a large number of R&D personnel. Companies that produce medicines or drug precursors routinely need a significant number of people in R&D, since that industry is constantly trying to produce new pharmaceuticals and other products.

In university settings, degrees are offered at the Associate, Bachelor, Masters, and PhD level, and all degree holders usually find employment in the industry if they actively seek it. Employment figures do show a downturn after the world recession of 2008, but chemists and chemical engineers did not seem to be as hard hit as other sectors of the economy. Employment in the chemical industry remains very steady.

1.5 Recycling

Recycling of materials has become a major concern in the past thirty years, and has spawned some companies devoted exclusively, or almost exclusively, to it. Plastics and metals come quickly to mind when one thinks of recycling, and it has been pointed out countless times that the recycling of such materials is both environmentally friendly and economically intelligent.

Paper and cardboard are often recycled as well. And while neither is generally considered a commodity chemical, there are enough chemical processes involved in the production of paper, paperboard, and cardboard that they deserve to be treated in any discussion of large scale recycling.

Whenever possible in each chapter, this book presents some discussion on the possibility of recycling the materials or chemicals that are the focus of that chapter. Perhaps obviously, there are some areas in which no recycling is done or can be done. When a chemical commodity is completely transformed, there is nothing left of the original material to recycle. When a material is entirely used by end-users, such as fertilizer, there is no way to re-gather any of it for recycling. In still other cases, the material is produced inexpensively on such a large scale, or is so long-lasting, that recycling is economically infeasible unless some as-of-yet unknown development occurs. But, whenever possible, when recycling takes place, it will be discussed in each chapter.

Bibliography

[1]   Arkema. (Accessed 7 February 2023, at www.arkema.com).
[2]   BASF. Website. (Accessed 7 February 2023, at www.basf.com).
[3]   Bayer. (Accessed 7 February 2023, at www.bayer.com).
[4]   Braskem. (Accessed 7 February 2023, at www.braskem.com.br).
[5]   Celanese. (Accessed 7 February 2023, at www.celanese.com).
[6]   Chemical & Engineering News. (Accessed 7 February 2023, at https://cen.acs.org/business/finance/
       CENs-Global-Top-50-2022/100/I26).
[7]   Degussa. (Accessed 7 February 2023, at www.degussa-goldhandel.de).
[8]   Dow. (Accessed 7 February 2023, at www.dow.com).
[9]   Dupont. (Accessed 7 February 2023, at www.dupont.com).
[10]   Eastman Chemical. (Accessed 7 February 2023, at www.eastman.com).
[11]   INEOS. (Accessed 7 February 2023, at www.ineos.com).
[12]   LyondellBasell. (Accessed 7 February 2023, at www.lyondellbasell.com).
[13]   Mitsubishi. (Accessed 7 February 2023, at www.m-chemical.co.jp).
[14]   Pittsburgh Plate Glass (PPG). (Accessed 7 February 2023, at www.ppg.com).
[15]   Saudi Basic Industries Corporation (SABIC). (Accessed 7 February 2023, at www.sabic.com).
[16]   Shell. (Accessed 7 February 2023, at www.shell.com).
[17]   Wanhua. (Accessed 7 February 2023, at https://en.whchem.com).

2 Sulfuric acid

2.1 Introduction

Sulfuric acid has been known and produced, on a relatively small scale, for several centuries. Today, sulfuric acid production has become one of the markers of a developed economy and of the robustness of a nation's economic health. This is because it is the largest commodity chemical produced on a world-wide scale in terms of volume. The basic raw material for it, sulfur, is obtained either from a mining process in which the element is extracted from underground deposits, or through reclamation of sulfides during ore purification, or through the desulfurization of crude petroleum, all of which are discussed below. The breadth of the extraction of sulfur, and thus the production of sulfuric acid from it, is evident from the USGS Mineral Commodities Summary, "In 2021, recovered elemental sulfur and byproduct sulfuric acid were produced at 95 operations in 27 States" [1]. This represents a remarkably large operation in just one of the world's developed nations.

2.2 Methods of production, sulfur

Both elemental sulfur and oxygen are required for the production of sulfuric acid, as well as water. The majority of sulfur has until recently been obtained from underground deposits, in an extractive well mining operation called the Frasch Process. The process utilizes three concentric tubes, which are inserted into the drill hole, into the sulfur deposit. The outermost tube is used for the injection of superheated water. Sulfur melts at 115 °C, and is forced to flow up through the middle tube. The innermost tube is used to force hot air into the molten sulfur-water mix, because the mix is more dense than water, and will not rise up with the water pressure alone. Thus, the hot air helps make a sulfur froth that can be forced up the middle tube. After drying, sulfur purity through this method of extraction can reach 99.7–99.8 %.

Sulfur can also be obtained as a by-product of metal refining from sulfide ores. As pollution controls for the refining of metals becomes more stringent, this method has become more economically viable. In this case, it is usually H_2S gas that is captured and used in the sulfuric acid synthesis, as hydrogen sulfide gas is the form taken by sulfides after roasting metal sulfide ores. The amount of sulfur captured in such a process becomes dependent upon the ore batch, as ores can be a mixture of metal oxides and sulfides.

In recent years, significant amounts of sulfur have been recovered from petroleum during its refining, in what is called the Claus Process. This is another form of sulfur capture that has become more attractive in recent years. This form of hydro-desulfurization of the heavier fractions of petroleum (the naphtha fraction, discussed in Chapter 9) provides sulfur, again from H_2S gas. The reaction is:

https://doi.org/10.1515/9783110671094-002

$$2H_2S + O_2 \longrightarrow 2S + 2H_2O$$

This method has taken a major place in the production of sulfur for use in the manufacture of the acid. While the driver for the increase ends up being economic, it is also a function of the allowable emissions from the refining, uses, and combustion of various petroleum sources and feeds.

2.3 Methods of production, sulfuric acid

The two major processes that produce sulfuric acid from sulfur are the Contact Process, and the Wet Sulfuric Acid Process. Small scale laboratory production is not discussed here.

Historically, sulfuric acid has been mentioned as far back as the tenth century. The Lead Chamber Process was used to produce it from the mid-1700s, which marks the beginning of industrial scale production, until early in the twentieth century, but has been displaced by the current processes, simply because these result in a more concentrated product. Lead Chamber Process acid tends to be isolated from 62–78 % purity.

2.3.1 The Contact Process

The reaction chemistry is essentially that shown in Scheme 2.1, below, a series of addition and redox reactions. Sulfur is first burned to form sulfur dioxide (SO_2). The second step, which forms sulfur trioxide, requires the use of a vanadium pentoxide catalyst (V_2O_5). This vanadium pentoxide catalyst has been an industry staple for almost a century. Catalyst lifetimes can be as high as 20 years. This reaction is very exothermic, and something of a testament to the robustness of the catalyst.

$$S + O_2 \longrightarrow SO_2$$
$$SO_2 + \frac{1}{2}O_2 \longrightarrow SO_3$$
$$SO_3 + H_2O \longrightarrow H_2SO_4$$
$$H_2SO_4 + SO_3 \longrightarrow H_2S_2O_7$$
$$H_2S_2O_7 + H_2O \longrightarrow 2H_2SO_4$$

Scheme 2.1: Sulfuric Acid Production, Reaction Chemistry.

When sulfur trioxide is formed, it is absorbed into existing 97–98 % sulfuric acid, resulting in a material known as fuming sulfuric acid (which still sometimes goes by the older name oleum), and which has the formula $H_2S_2O_7$. The final step is the dilution of the fuming sulfuric acid to produce concentrated sulfuric acid. Direct addition of sulfur trioxide to water is not done on a large scale, because it is very exothermic and forms a sulfuric acid aerosol that is very corrosive, and hard to capture and use.

2.3.2 The Wet Sulfuric Acid Process

This process has become common in the past twenty years, and has proven to be a highly efficient means of reducing sulfur emissions from gases. Additionally, because the reactions are highly exothermic, steam is always produced along with the acid. The steam can also be used, often for heating.

The reaction chemistry for the Wet Sulfuric Acid Process is summed up in Scheme 2.2, below:

$$H_2S + 3/2O_2 \longrightarrow SO_2 + H_2O$$

$$SO_2 + \frac{1}{2}O_2 \longrightarrow SO_3$$

$$SO_3 + H_2O \longrightarrow H_2SO_{4(g)}$$

$$H_2SO_{4(g)} \longrightarrow H_2SO_{4(l)}$$

Scheme 2.2: Wet Sulfuric Acid Process Reaction Chemistry.

Alternatively to the reactions listed above, hydrogen sulfide can be used as the sulfur source. The reaction is exothermic, and is often expressed as follows (like the above):

$$2H_2S + 3O_2 \longrightarrow 2H_2O + 2SO_2$$

This step produces roughly 10 % SO_2, all of which is then fed into a converter using a V_2O_5 catalyst for conversion to SO_3 (roughly 60 % conversion), with a short contact time of 2–4 seconds.

2.4 Volume of production and sales

Sulfuric acid is produced on a scale of hundreds of millions of tons annually. The fluctuations in its output are usually tied directly to the swings of the world's economy. The opening or closing of a metal refinery can also affect world production of sulfuric acid, because metals are often extracted from sulfide ores. The sulfur from the ore is used in such cases as a source for producing the acid.

Sulfuric acid continues to be sold as a bulk chemical at approximately $146 per ton. This translates to just less than $0.073 per pound, and has fluctuated only marginally in the past decade.

2.5 Uses

2.5.1 Production of phosphoric acid

The largest single use for sulfuric acid remains its consumption in the production of phosphoric acid, which is ultimately used for the production of phosphate-based fertilizers. The reaction chemistry is as follows:

$$Ca_5F(PO_4)_3 + 5H_2SO_4 + 10H_2O \longrightarrow 5CaSO_4 \cdot 2H_2O + 3H_3PO_4 + HF$$

While this provides enormous amounts of phosphate for fertilizers, it also produces calcium sulfate, which forms as a precipitate, and tends to concentrate impurities from the phosphate rock, including actinide impurities.

Phosphoric acid is also used in a variety of applications besides fertilizer. Common applications include:

- Household cleaner. This is found in many cleaning materials as sodium salts of the acid.
- Food additive (high grade). This is often used to lower the pH of certain colas and provide a slightly sour taste.
- Dispersant. Phosphoric acid is highly polar and soluble in water, and thus can be used as a dispersant.
- Rust inhibitor. H_3PO_4 converts iron (III) oxide into iron (III) phosphate, which can be removed from an iron surface, generating a new surface.
- There are also a number of smaller volume uses.

2.5.2 Production of aluminum sulfate

Aluminum sulfate, sometimes still referred to by the older name 'alum,' can be produced according to the following reaction:

$$2Al(OH)_3 + 3H_2SO_4 \longrightarrow Al_2(SO_4)_3 + 6H_2O$$

The ultimate source of the aluminum is bauxite. Aluminum refining is discussed later, in Chapter 24. Since the sulfate is routinely isolated as a hydrate, the reaction can be more properly written:

$$2Al(OH)_3 + 3H_2SO_4 \longrightarrow Al_2(SO_4)_3 \cdot 6H_2O$$

Aluminum sulfate can be used as a flocculating agent to aid in the purification of water for human consumption. It also can be found in the following areas:

1. Textile printing
2. Dyeing and printing

3. Soil pH adjuster
4. Baking powder (a component)
5. Antiperspirant ingredient
6. Fire-fighting foams, when added to sodium bicarbonate

2.6 Recycling

Ironically, the continued low price for sulfuric acid works against any need or desire to recycle it, as does its use and reactivity in making further products. Sulfuric acid that becomes part of any consumer product – automobile batteries, for example – is usually neutralized and discarded, even though the lead and the plastic casing for such batteries are recycled. This is because such small amounts of sulfuric acid are used per battery that there is no economic incentive to recover it.

However, sulfuric acid that is used in the petroleum refining industry can be reclaimed. Between 2.5 million and 5 million tons of the acid per year ends up being recycled in this manner [2]. The reason this recovery and recycling is possible is because the material is not widely distributed into small, end user applications, but is localized to the industrial sites.

Likewise, the products we have discussed that are produced from sulfuric acid are usually manufactured on a large enough scale that they are not recovered, or they are used in consumer end products, and are thus not recovered.

Bibliography

[1] Mineral Commodity Summaries 2022, US Department of the Interior, U. S. Geological Survey, ISBN 978-1-4113-4434-1.
[2] Sulfuric Acid Today. Trade magazine for the industry, published twice annually. Website. (Accessed 13 September 2012, at www.h2so4today.com/).

3 Major industrial gases

3.1 Introduction, air liquefaction

The three gases, oxygen, nitrogen, and argon are all extracted from a natural, essentially limitless source – air. The process is air liquefaction, and the separation is ultimately dependent upon the different boiling points of the three different elements.

The other noble gases, with the exception of helium, are all also extracted from air during the liquefaction process. The heavier noble gases are all present in air in very small amounts. Helium is still generally obtained worldwide as a by-product of natural gas well exploitation.

3.2 Helium

Although helium is not used in quantities as large as nitrogen or oxygen, it does have several important scientific and consumer applications. It is currently extracted from underground natural gas deposits in Qatar, Algeria, Iran, and Russia, as well as the United States, and separated cryogenically. Estimates for helium reserves are tabulated by the United States Geological Survey. Helium resources of the world, exclusive of the United States, were estimated to be about 31.3 billion cubic meters (1.13 trillion cubic feet). The locations and volumes of the major deposits, in billion cubic meters, are Qatar, 10.1; Algeria, 8.2; Russia, 6.8; Canada, 2.0; and China, 1.1 [1].

The United States Texas-Oklahoma western panhandle, and the Colorado and Wyoming helium facilities run by Exxon Mobil currently account for approximately 25 % of US use [4].

3.3 Helium uses

While many people think of party balloons as the major consumer end product that requires helium, there are a number of different uses, some requiring higher purity than others, for helium in scientific and business applications. They are listed below, in Table 3.1. As hospitals open new MRI clinics, there will be a greater need for liquid helium, and stresses will continue to be placed on existing sources. While recycling and recovery are in their early stages, several larger research universities and institutes are examining the purchase of one or more gaseous helium recovery systems, as a cost saving measure when compared to the continued purchase of liquid helium for high field NMRs.

https://doi.org/10.1515/9783110671094-003

Table 3.1: US Consumption of helium in 2021, in declining amounts (40 Mm^3 total).

Helium application
Magnetic resonance imaging
Lifting gas
Analytical and laboratory applications
Electronics and semiconductor manufacturing
Welding
Engineering and scientific applications
Various other (minor) applications

3.4 Air liquefaction

The two methods that can be used to liquefy air are the Linde Process, and the Claude's Process. Both are engineering processes, and not chemical reactions that effect the separation.

The first involves the repeated compression, cooling, and expansion of the air sample, with the result that the air, when expanded, is always colder than at the beginning of the cycle. The end result is air that is cool enough that it becomes liquid. Starting pressure for the air mixture is usually 60 psi or greater, and the end is an oxygen-enhanced liquid phase and nitrogen.

The latter process is quicker, involves a double chamber and the Joule–Thomson Effect as the gas temperature decreases to the point of liquefaction.

Both processes depend upon the different boiling points of nitrogen and oxygen, as well as the different boiling point of argon, to effect the separations of the gases, as seen in Table 3.2. The major cost of the process over the long term is that of the energy needed to cool the mixture – the air – with a series of compressors.

Table 3.2: Boiling Points of Air Components.

Component	Boiling Point (in K)
Oxygen	90.2
Argon	87.3
Nitrogen	77.4

The separated components can be transported for short distances via pipeline, if there is a large scale need. Otherwise, Dewar flasks and compressed gas bottles are used for distribution.

3.5 Oxygen

3.5.1 Storage and uses

Bottled oxygen can be stored as a compressed gas in steel cylinders, but liquefied oxygen can be stored in Dewar containers, which come in a wide variety of sizes.

Somewhat more than half of all liquefied, separated oxygen is used to smelt iron ore into steel. Oxygen is injected into the molten iron in order to react with sulfur impurities as well as carbon impurities. These impurities are then removed as oxides, generally SO_2 and CO_2. Pressurized oxygen, once returned to the gas state, is often used in industrial cutting and welding of metals, as it raises the working temperature of an acetylene cutting torch.

In general, large uses of oxygen include:
- Iron refining.
- Ethylene production.
- Metal cutting/welding.
- Medical applications.

Since these applications are very common in the industrialized world, purified oxygen is required and used in almost all nations. While there are several large industrial firms that produce oxygen, all produce other gases as well. Air Liquide states at its website that: "Air Liquide is a world leader in gases, technologies and services for Industry and Health. Oxygen, nitrogen and hydrogen are essential small molecules for life, matter and energy. They embody Air Liquide's scientific territory and have been at the core of the company's activities since its creation in 1902" [2]. While other producers may not be as large a corporate entity, all that produce oxygen produce at least nitrogen in addition [3, 6, 5].

3.6 Nitrogen

3.6.1 Major uses

By far, captured nitrogen's largest use is the production of ammonia fertilizer. This is so large that the USGS Mineral Commodity Summaries uses the word "Nitrogen" in its annual publication to mean ammonia – what they term 'fixed nitrogen.'

Since the advent of the Haber Process in the early part of the 20th century, the production of fertilizer directly from nitrogen and hydrogen at elevated pressure in a catalyzed system has replaced any natural source of nitrogen-containing fertilizer. This will be treated in detail in the next chapter.

3.7 Argon

3.7.1 Major uses

Approximately 1% of the lower Earth's atmosphere is argon, and thus the production of argon becomes the third product in air liquefaction plants, as we have seen. Argon has no industrial use that involves reaction chemistry. Rather, almost all of it is used in producing inert atmospheres for a variety of applications.

3.8 Carbon dioxide

Production of carbon dioxide simply requires a carbon source, usually in a gas phase material. Natural gas is often stripped of its hydrogen, which then leaves carbon dioxide gas. The gas can be refrigerated and pressurized (CO_2 liquefies at <6 atm), resulting in liquid CO_2. This is followed by quick pressure reduction, which vaporizes a portion of the material – which in turn cools the remainder enough that it solidifies. The resulting solid CO_2 can then be formed into blocks or pellet-sized units.

3.8.1 Carbon dioxide uses

The following are the major uses of carbon dioxide, often as dry ice:
- Cooling, storage, and preservation of food
- Carbonating beverages.
- Blast cleaning.
- Enhanced oil recovery – injection of CO_2 into the well forces oil from rock formations.
- Inert atmosphere for grain storage.
- Controlling the pH of wastewater streams.

In addition, small amounts of carbon dioxide are used in several consumer applications, including:
- Theater effects.
- Fog machines in night clubs, theaters, and haunted houses.
- De-gassing tanks used for volatile, flammable chemical storage, although argon can be used in this manner as well.

While dry ice is considerably colder than ambient temperatures in most work areas, it can be stored for days or weeks in standard freezers, although it does undergo slow sublimation in such environments.

3.9 Pollution, recycling, by-product uses

The pollution generated by an air liquefaction plant is essentially that of its power signature. The materials, air, and the products, the separated, liquefied elements, are not hazardous.

Recycling in the traditional sense is not considered possible, because none of the materials are captured. Nitrogen, when used as an inert atmosphere, is simply released into the surrounding atmosphere. Oxygen, when used as an element in medical or other applications, is similarly released. All uses of argon similarly release it to the surrounding air. Only recently has the price of helium risen to a point where the recovery of gaseous, bled-off helium from nuclear magnetic resonance spectrometers has become an option for universities and research centers [7].

The by-products of air liquefaction are the heavier noble gases. The overall volume of these however is small enough that there are currently no recycling programs for them of any significant size.

Bibliography

[1] USGS Mineral Commodity Summaries, 2022. (Accessed 21 September 2022, downloadable).
[2] Air Liquide. Website. (Accessed 21 September 2022, at www.us.airliquide.com/).
[3] Air Products. Website. (Accessed 21 September 2022, at www.airproducts.com/).
[4] "Helium Supplies are Scant, Again," Chemical & Engineering News, 16 July, 2012, pp. 32–34.
[5] Emsley, John (2001) "Oxygen". In: Nature's Building Blocks: An A–Z Guide to the Elements, Oxford, England, UK: Oxford University Press, pp. 297–304. ISBN 0-19-850340-7.
[6] Linde. Website. (Accessed 21 September 2022, at www.linde.com).
[7] The Impact of Selling the Federal Helium Reserve. National Academies Press. (Accessed 21 September 2022, as: https://nap.nationalacademies.org/read/9860/chapter/7).

4 Nitrogen compounds

4.1 Inorganic nitrogen compounds

We have seen how nitrogen is extracted from air in Chapter 3. One of nitrogen's main uses is the manufacture of ammonia via the Haber Process. This chapter will examine both the production of ammonia, and the uses of it to produce the other large scale nitrogen-containing compounds: ammonium nitrate, nitric acid, urea, and ammonium sulfate.

4.2 Ammonia

4.2.1 Introduction and history

Nitrogen-containing fertilizers have been used throughout history, but synthetic ammonia has only been available for this application for just over the last century. Natural sources have routinely incorporated animal wastes in some form or another. The Haber Process, named for the German chemist Fritz Haber, was first unveiled in 1909, but was initially too slow to produce ammonia on an industrial scale. The process was purchased by BASF, where chemist Carl Bosch was assigned the project of scaling the process up to a scale that would make it an industrially profitable material. The problems had been solved by 1913, the first year in which ammonia was produced industrially.

4.2.2 Sources of starting materials

We have discussed the liquefaction of air in the previous chapter, and will explore the hydrocarbon stripping that produces hydrogen in Chapter 10. For the moment, we will note that the amount of hydrogen required for industrial scale production of ammonia is currently only obtainable from the methane obtained from natural gas.

4.2.3 Reaction chemistry

The reaction chemistry for the production of ammonia appears to be remarkably simple, although how it was derived remains one of the great chemical accomplishments of the early twentieth century.

$$3H_{2(g)} + N_{2(g)} \longrightarrow 2NH_{3(g)}$$

What is usually omitted from this reaction is the set of conditions under which it runs, and the source of hydrogen (which was just mentioned).

https://doi.org/10.1515/9783110671094-004

Hydrogen and nitrogen are mixed at 400–600 °C in the presence of an iron oxide catalyst. After the mixing, the product is cooled which condenses part of the ammonia as a liquid. Recycling the remaining gaseous material increases the overall yield from what would be approximately 20 % for one contact cycle to 85–90 % overall.

4.2.4 Uses

As mentioned, a large portion of ammonia is used directly as fertilizer [3], while a large portion of the remainder is converted into a few solid derivatives, which are also largely used as fertilizer.

4.3 Nitric acid

Nitric acid is often sold in commercial grades that are quite low in purity, when compared to many other bulk chemicals (roughly 52–68 % nitric acid). The Ostwald Process is used to manufacture nitric acid on a large scale. The reaction chemistry is:

$$4NH_3(g) + 5O_2(g) \longrightarrow 4NO(g) + 6H_2O(g) \quad \Delta H = -905\,kJ$$

Then

$$2NO(g) + O_2(g) \longrightarrow 2NO_2(g) \quad \Delta H = -114\,kJ$$

Then

$$3NO_2(g) + H_2O(l) \longrightarrow 2HNO_3(aq) + NO(g) \quad \Delta H = -117\,kJ$$

If the final step is carried out in the ambient environment, in air:

$$4NO_2(g) + O_2(g) + 2H_2O(l) \longrightarrow 4HNO_3(aq)$$

All of which are exothermic. In the first step, ammonia is oxidized to nitrogen monoxide using a platinum or rhodium catalyst at approximately 500 K (220 °C) and a 9 bar pressure. The second major step involves reacting nitrogen monoxide with air to further oxidize the nitrogen. The final step is the absorption of nitrogen dioxide in water to form the product and more nitrogen monoxide. This by-product is then recycled into the system to increase the overall yield of nitric acid.

Beyond this, nitric acid can be concentrated to 98 % when the final product is dehydrated with sulfuric acid.

4.4 Ammonium nitrate

Ammonium nitrate is produced in a straightforward manner from nitric acid and ammonia. A variety of different conditions can be used to manufacture a product in the form of prills or various types of crystals [1].

$$NH_3 + HNO_3 \longrightarrow NH_4NO_3$$

4.5 Adipic acid

The production of adipic acid, used in the synthesis of nylon, is made from cyclohexanol and nitric acid. The reaction chemistry, somewhat simplified is as follows:

$$HNO_3 + C_6H_{11}OH \longrightarrow HNO_2 + H_2O + C_6H_{10}O$$
$$C_6H_{10}O + HNO_3 \longrightarrow HO_2C(CH_2)_4CO_2H$$

A more detailed discussion of nylon is deferred until Chapter 15, Polymers.

4.6 Urea

While urea is now produced industrially, throughout history it was extracted from animal wastes, and used as a form of fertilizer. It was Friedrich Woehler who first synthesized urea from an inorganic source, in 1828, perhaps inadvertently helping to delineate what would later be called organic chemistry from inorganic chemistry.

The production of urea requires ammonia and carbon dioxide at elevated temperature and pressure.

$$2NH_3 + CO_2 \longrightarrow NH_2COO^-NH_4^+ \longrightarrow NH_2CONH_2 + H_2O$$

The carbon dioxide produced during hydrocarbon stripping of natural gas can be used for the production of urea. The process by which urea is produced is now called the Bosch–Meiser Urea Process, and has been in use since the early 1920s. The first step utilizes CO_2 as dry ice, and produces ammonium carbamate. The second step is a decomposition that produces the urea. Taken together, the entire process is an exothermic one.

Over 90 % of urea is used as fertilizer, largely because it has a high nitrogen content for its molecular mass.

4.7 Ammonium sulfate

Ammonium sulfate is yet another important commodity chemical in the fertilizer industry [2]. Both the nitrogen and the sulfur are useful for plant growth, and ammonium sulfate contains 21 % by weight of nitrogen. It is often used in slightly basic, or alkaline, soils, because the ammonium release makes the soil more acidic.

$(NH_4)_2SO_4$ is generally manufactured through the direct combination of ammonia with sulfuric acid. The reaction chemistry can be represented simply, as seen here:

$$H_2SO_4 + 2NH_3 \longrightarrow (NH_4)_2SO_4$$

The reaction proceeds by the introduction of water vapor and gaseous ammonia, and the actual combination of this with free sulfuric acid is exothermic enough that the solution often remains at or near 60 °C. The solid product is recovered from the evaporation of the water from the system.

While there are uses for ammonium sulfate beyond fertilizers, it is again the fertilizer industry that consumes the largest portion of it [4]. Food grade ammonium sulfate is sometimes used in bread and flours to regulate the overall acidity of the end product.

In recent years, the import of ammonium sulfate has been banned in some areas of Afghanistan and Pakistan, because of fears that it is being used for the manufacture of explosives.

4.7.1 Pollution

Much of the pollution associated with ammonia use is that involved in the runoff of ammonia fertilizers when an excess is applied to farm fields, and not from the actual manufacture of the material. Runoff gathers from small streams to larger rivers, ultimately affecting such large, inland waterways as the Rhine, the Ohio, the Missouri and the Mississippi.

Nitric acid is produced from elemental materials that use air as a feedstock, and natural gas for the hydrogen source. Thus, it would seem to be a material that has a low pollution footprint. However, the escape of any nitrogen oxides into the atmosphere, at any step in the production, remains a cause of concern.

4.7.2 Recycling possibilities

The idea of recycling ammonia is difficult, in part because bulk prices have remained consistently low for decades, but mostly because the end use of the material usually distributes it over wide, often rural areas. There are no major ammonia recycling programs currently in effect in the industrialized world.

Likewise, nitric acid is almost always used in some form of reaction chemistry, making recycling of this commodity very difficult.

Bibliography

[1] Ammonium Nitrate Nitric Acid Producers Group. Website. (Accessed 22 September 2022, at www.an-na.org/).
[2] The Fertilizer Institute. Website. (Accessed 22 September 2022, at www.tfi.org/).
[3] International Fertilizer Industry Association. Website. (Accessed 22 September 2022, at www.fertilizer.org).
[4] Nitrogen and Syngas, a trade magazine. Website. (Accessed 22 September 2022, at www.bcinsight.com/publication.asp?pubID=1).

5 Chemicals from limestone

5.1 Introduction

Limestone, $CaCO_3$, has been a major building material for centuries, and in its most visually attractive form – marble – has been used to carve statuary and sculpture for just as long. Much of the world's limestone is composed of the fossilized remains of ancient coral and marine organisms. Limestone is considered, however, to be a sedimentary form of rock.

In the latter half of the nineteenth century, the Solvay Process was developed and scaled up to an industrial process by which limestone and sodium chloride are converted into soda ash (Na_2CO_3) and calcium chloride.

Sources of limestone are numerous and large. It is quarried or mined in many areas of Europe and North America, making it a plentiful feedstock for the production of other chemicals, often calcium-containing materials. Some well-known areas are almost entirely made of limestone, such as the Florida Keys. Certain limestone deposits have impurities which actually make the stone more visually attractive, and thus preferable to other materials in certain decorative building applications. The whitest marble is routinely composed of calcium carbonate that was deposited in a formation that was very low in silicate minerals [2].

Since limestone or the basic chemicals produced from it are used in such a large array of products and industries, it is tracked by the United States Geological Survey [3], as shown in Table 5.1.

Table 5.1: Uses of Limestone.

Use	Description
Building material	Stone buildings and structures
Agriculture	pH adjustment and neutralization of acidic soil
Fluxing	Iron extraction from ore (discussed in Chapter 23)
Sulfur removal	Reacts with SO_2 in flue gas
Road base	Crushed stone used below road surfaces
Marble	For decorative carving and sculptures
Food grade	Added to foods as a calcium source
White pigment	In foods and paints

5.2 Lime

Lime is produced from limestone by heating calcium carbonate at 1200–1300 °C. This requires large amounts of fuel to maintain the required heat, although the reaction chemistry appears quite simple, as follows:

https://doi.org/10.1515/9783110671094-005

$$CaCO_3 \longrightarrow CaO + CO_2$$

The reaction is performed on a millions of tons per year scale, and tends to be a multi-billion dollar per year process. Companies manufacture lime on all six continents, and in the developing as well as the developed world [3].

The names used for both the starting and end material sometimes still follow older conventions, and do not always agree with chemical nomenclature. Table 5.2 illustrates much of the different nomenclature.

Table 5.2: Names for calcium-containing materials derived from limestone.

Formula	Nomenclature	Common Names
$CaCO_3$	Calcium carbonate	Limestone, calcite, marble, chalk
CaO	Calcium oxide	Lime, unslaked lime, quicklime
$Ca(OH)_2$	Calcium hydroxide	Slaked lime, hydrated lime

The USGS lists the uses of lime in descending order as: "steelmaking, flue gas desulfurization, construction, water treatment, mining, precipitated calcium carbonate, and pulp and paper" [3]. The use of lime in the paper industry will be discussed in Chapter 18.

5.3 Sodium carbonate

The production of sodium carbonate through what is called the Solvay Process is one of the major uses of limestone and of sodium chloride [1, 3, 4, 5]. The reaction chemistry can be written succinctly, as follows:

$$CaCO_3 + 2NaCl \longrightarrow Na_2CO_3 + CaCl_2$$

But the full process is considerably more complex. The following are the broad steps by which soda ash is produced:

$$NaCl + CO_2 + NH_3 + H_2O \longrightarrow NaHCO_3 + NH_4Cl$$

In this step, carbon dioxide is passed through an ammoniated brine solution, often in towers as high as 80 ft. The bicarbonate then precipitates because of its basicity, and almost all of the ammonia is recovered and recycled for further use.

$$CaCO_3 \longrightarrow CaO + CO_2$$

The heating of limestone produces the CO_2 that is used in the process. This is an energy intensive step, usually carried out at 1,000–1,100 °C.

$$CaO + 2NH_4Cl \longrightarrow CaCl_2 + 2NH_3 + H_2O$$

The calcium oxide, or quicklime, is also used after its formation from calcium carbonate, by being reacted with the hot ammonium chloride that has been produced, after the sodium bicarbonate has been removed. The ammonia is, as mentioned, recycled.

$$2NaHCO_3 \longrightarrow H_2O + CO_2 + Na_2CO_3$$

This final reaction is run at 160–230 °C, produces the final product, and also produces CO_2 which is recovered and recycled into the process.

It can be seen from these four reactions that while sodium carbonate or sodium bicarbonate can be recovered, the other materials are all recycled into the process – with the exception of calcium chloride. This by-product of the Solvay Process has been of interest to sodium carbonate producers for decades, simply because it is produced in large enough quantities that a market designed specifically for it would be highly lucrative.

5.3.1 Soda ash uses

Soda ash is produced on this large a scale simply because there is a wide variety of uses for it, some of them requiring great quantities. In broad terms, soda ash use can be divided as follows, in Figure 5.1.

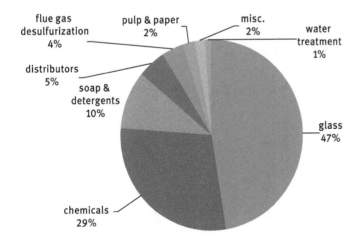

Figure 5.1: Soda Ash Use by Commercial Sector [2].

5.3.2 Sodium bicarbonate uses

Sodium bicarbonate has extensive uses in the following areas: food products such as baking soda, toothpastes and other personal hygiene areas, fire extinguishers, as well as in the pharmaceutical markets.

Since sodium bicarbonate produced in the Solvay Process is usually isolated at approximately 75 % purity, it is economically more feasible to react the product, sodium carbonate, with water and CO_2 to produce the bicarbonate at a high enough grade that it can be used for human consumption.

5.4 Calcium chloride

Calcium chloride remains the one major material produced in the Solvay Process that really had no major, intended use, at least not at the beginning of the industrial-scale execution of the process. Thus, the price for calcium chloride has consistently remained low, and there have been a multitude of people who have looked for uses and applications for this material. It is now used as a road de-icer, and competes with sodium chloride in this regard. But the accumulation of calcium chloride at Solvay plants that can not discharge waste solution to an ocean has meant severe pollution in some lake areas. Discharge of excess calcium chloride into the ocean has not appeared to have caused environmental damage, at least thus far.

5.4.1 Uses

Calcium chloride has found numerous uses, on both large and small scales. These include the following:
– Road de-icing material.
– Brine, for refrigeration.
– A desiccant, which has been approved for food use by the US Food and Drug Administration.
– A firming agent, in various canned foods.
– A milk and beer additive, as a way to adjust the ion content of the solution (and thus better the consumable product).

But all of these uses post-date the advent of the Solvay Process, and are essentially ways to develop uses for a material that was a by-product of sodium carbonate production.

Interestingly, calcium chloride is also used to decrease the melting point in the production of sodium metal. The process is essentially the electrolysis of molten sodium chloride (seen in Chapter 6), and the result is sodium metal. Sodium production, while large, remains a relatively small user of calcium chloride.

5.5 Pollution and recycling

The Solvay Process is one of the largest in the world, and it recycles all its reactants and materials – except for one. As mentioned, calcium chloride is produced as a by-product, and is the second product when ammonia is recycled from ammonium chloride. As mentioned, because the Solvay Process produces so much sodium carbonate on a world wide scale, uses for calcium chloride have been thought up, discovered, and marketed, even though they are not strictly necessary. For example, large amounts of calcium chloride are sold as road de-icer, sometimes with the claim that on a per particle basis, the material de-ices 50 % more than traditional sodium chloride. Anyone who has been in the less than ideal situation of driving behind a salt truck in the winter, especially when it continues to spew a torrent of salt while idling at a traffic light, might have wondered if the driver has ever heard this. Thus, calcium chloride becomes a material deposited on roads, which ultimately ends up near and at the side of roads. Excesses of it can make its way into storm drains and local waters.

Based on the starting feed for any Solvay chemistry, certain amounts of magnesium and calcium will be present. These ions are routinely precipitated out as carbonates prior to a batch of feedstock being reacted. Thus, these become a secondary form of waste product for the Solvay Process.

Bibliography

[1] European Soda Ash producers Association. Website. (Accessed 24 September 2022, at www.esapa.eu/).
[2] Natural Stone Institute. Website. (Accessed 24 September 2022, at: https://www.naturalstoneinstitute. org).
[3] Mineral Commodity Summaries 2022, US Department of the Interior, U. S. Geological Survey. (Accessed 24 September 2022, downloadable).
[4] Solvay. Website. (Accessed 24 September 2022, at www.solvay.com/en/).
[5] Solvay Chemicals North America. Website. (Accessed 24 September 2022, at www.solvay.com/en/usa).

6 Sodium chloride

6.1 Introduction

Salt has been known to humankind and used in many ways since ancient times. It has been associated with good health for millennia, and has been used as a food preservative for almost all of that time.

At times salt has been used as a medium of exchange. The expression, "to be worth one's salt" has its origin in this. Roman soldiers were at times paid in salt, and the word "salary" also has its origin in this practice.

When Jesus states in the Gospel of Matthew, "You are the salt of the earth. But if salt loses its taste, with what can it be seasoned? It is no longer good for anything but to be thrown out and trampled under foot," he is referring to the salt used by the poorer people, which was often mined and thus had a certain amount of dirt and clay in it. When such salt was significantly contaminated with dirt, it had to be discarded. People at the time were unaware that such salt could be dissolved in water, separated from the clay or dirt, then recrystallized as a pure material by evaporation of the water.

In addition to the health benefits and the preservative uses of salt, the poisonous nature of it in and on the earth has been known for millennia, which is why Roman legions salted the earth in and around Carthage to celebrate their victory over that city-state. By doing so, they ensured the fields around Carthage would never be fertile again, and that their victory would thus be complete. The fact that they had enough salt to use it as a weapon indicates how long salt has been produced and utilized as a large-scale commodity.

Today, salt is one of a very few commodities from which an enormous number of materials are manufactured. The chloride in polyvinyl chloride comes from salt. The sodium used as a moderator in nuclear reactors likewise has its origin in salt. While some materials that come ultimately from salt will be covered elsewhere in this book, we have grouped several of the large commodity chemicals in this chapter, all of which require salt for their production.

6.2 Methods of recovery, production, and extraction

Production of salt is common in many countries, either from mines or through extraction from the sea [5]. Most countries and corporations consider the supply of salt inexhaustible, although some land-locked countries may at times be cut off from sources of it. The US Geologic Survey keeps track of and tabulates salt production each year, and does delineate between open pan salt, solar salt, and rock salt, the names reflecting how the material is isolated. Recent statistics are listed in Table 6.1, and Figure 6.1, below.

https://doi.org/10.1515/9783110671094-006

Table 6.1: Salt Production by Country, (in thousands of metric tons) [4].

Country	Production 2010	Production 2011 (est.)
Australia	11,968	13,000
The Bahamas	10,000	10,000
Brazil	7,020	7,000
Canada	10,537	11,000
Chile	8,400	9,000
China	62,750	65,000
France	6,100	6,000
Germany	19,100	20,000
India	17,000	18,000
Mexico	8,431	8,800
The Netherlands	5,000	5,000
Pakistan	11,000	11,000
Poland	3,520	4,000
Spain	4,350	4,400
Turkey	4,000	4,000
Ukraine	5,400	5,500
United Kingdom	5,800	5,800
United States of America	43,300	44,000
Other countries	36,200	39,000
World total	280,000	290,000

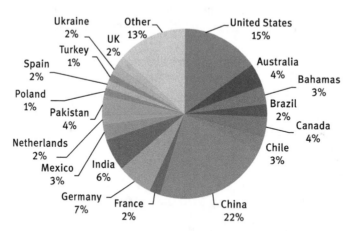

Figure 6.1: National Salt Production, as a percentage of the total [4].

6.2.1 Evaporation

Salt can be mined through a brine solution process, in which hot water is injected into the earth, then hot brine solution is extracted. The brine solution is then evaporated to produce salt in high purity, because contaminant ions, such as Ca^{2+} and Mg^{2+} can be precipitated out selectively, as the carbonates, leaving only sodium chloride. Salt purity in this process can be greater than 99 %.

Additionally, the Grainer evaporation process – an open pan process – can be used to produce crystalline salt from a brine solution. The process is energy intensive because it requires immersion coils in the brine to heat it, and may require a partial vacuum to help speed the evaporation process.

6.2.2 Mining

Rock salt can also be mined directly from underground deposits, without forming a brine solution. The product is generally referred to as halite, and is not always of as high a purity as evaporated salt, at least not before treatment to remove other, naturally included ions. Again, such ions are normally removed through solvation, concentration and precipitation, then evaporation of water.

6.2.3 Uses

The largest single use of sodium chloride is the production of sodium hydroxide (caustic soda) and chlorine gas, both of which are discussed below [2, 6]. The common image of salt being used to de-ice roads and highways does account for close to 20 % of salt use, but calcium chloride, and to a lesser extent, wood ashes and sand, compete with it in this application. Wood ashes and sand are now considered a somewhat more environmentally friendly option for road de-icing. As well, the de-icing use is not all that much larger than the uses of salt in the food industry, both for humans and farm animals. There are also a variety of smaller, specialty uses for salt.

Salt used for direct human consumption is almost always iodized, meaning iodine is added directly to the product prior to sale. This prevents the development of goiter, a disease affecting the thyroid.

6.3 Chemicals produced from sodium chloride

6.3.1 Sodium hydroxide production

The largest industrial use for sodium chloride is the production of sodium hydroxide, sometimes called industrial caustic, or lye. The secondary products of the reaction are chlorine and hydrogen gas. Most hydrogen gas that is used as a separate starting material for further reaction chemistry is produced from the hydrocarbon stripping of the lightest fraction of crude oil or natural gas, and will be addressed in Chapter 10. The Eurochlor website states, "more and more companies use their excess hydrogen in fuel cells, to generate electric power" [3]. Thus, this becomes the use for hydrogen captured in sodium hydroxide production.

The reaction chemistry for sodium hydroxide production is a fairly straightforward oxidation–reduction reaction, and is shown below:

$$2NaCl_{(aq)} + 2H_2O_{(l)} \longrightarrow 2NaOH_{(aq)} + H_{2(g)} + Cl_{2(g)}$$

The process is somewhat more complicated than the reaction chemistry implies, and there are three separate ways in which sodium hydroxide is manufactured. All three methods are considered to be variations of the Chlor-Alkali Manufacturing Process.

6.3.2 The mercury cell

In this type of cell, the cathode is a flowing layer of elemental mercury in which sodium is actually amalgamated. This functions as the cathode. In a separate chamber, after the oxidation of chloride to chlorine gas, the sodium must be re-oxidized while water is split and combined with it, the reduced hydrogen being taken off as the elemental gas.

This type of operation is capable of producing a solution of 50 % caustic.

Because of the toxicity of elemental mercury, the use of this type of cell has declined in the recent past. In the United States, leaders in the Chlor-Alkali Process industry are working with the Environmental Protection Agency to continue to reduce mercury emissions from these plants. A larger number of plants in Europe utilize this cell design.

6.3.3 The diaphragm cell

The diaphragm cell is now the major method by which the Chlor-Alkali Process is executed, as seen in Figure 6.2. In it, the anode and cathode are separated by a diaphragm made of asbestos with Teflon mixed into the diaphragm. This surrounds and coats each cathode in a cell, and controls migration of the electrolyte solution to the cathode.

6.3.4 The membrane cell

This cell design is still considered the newest of the three, and replaces the asbestos diaphragm with an ion exchange membrane. Currently, almost one third of the Chlor-Alkali plants utilize this design. The advantages are elimination of any mercury emission, and the elimination of the use of asbestos.

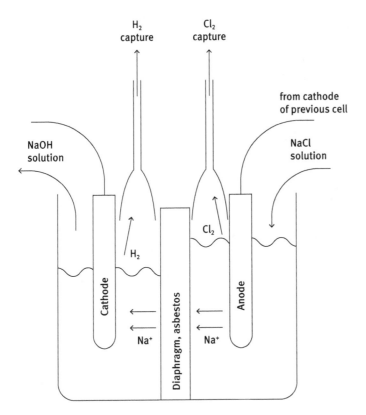

Figure 6.2: General Scheme of the Chlor-Alkali Process.

6.3.5 Production statistics

There are several major producers of sodium hydroxide, a number of which are multinational companies. The list includes:
1. Dow Chemical Company, in Freeport, Texas, and in Plaquemine, Louisiana.
2. Oxychem.
3. Pittsburgh Plate Glass.
4. Olin.
5. Pioneer Companies, Inc. (PIONA, which has been purchased by Olin).
6. Formosa.

But there are others, worldwide, as well. These six are major producers.

Statistics for the production of sodium hydroxide and chlorine are maintained by the Chlorine Institute [2], and updated monthly. The Chlorine Institute website states, "the Chlorine Institute offers a monthly report on industry production capacity for chlorine, and the quantity of chlorine actually produced, liquefied and shipped for each month. It also includes production data for sodium hydroxide" [2].

6.4 Uses of NaOH

According to the World Chlorine Council, worldwide production of sodium hydroxide has risen to 54 million metric tons [6]. The uses of sodium hydroxide are many, and in many cases the material becomes an intermediate for other end use products. A partial list is shown in Table 6.2, below [6, 3]:

Table 6.2: Uses of Sodium Hydroxide and Derivatives.

NaOH Derivative	Intermediate?	End Product
Acrylonitrile		ABS resins, household and automotive plastic parts
Sodium cyanide	Adiponitrile	Nylon
Sodium formate		Rare earth element processing
Sodium chlorite		Bleach in textiles
Sodium laurel sulfate		Food additive
Sodium stearate	Aluminum/calcium/zinc stearate	Gelling agent, cosmetics, soaps
Monosodium glutamate		Food flavor enhancer

6.5 Chlorine

One of the biggest changes in human health in history occurred at the beginning of the 20th century, in the isolation of elemental chlorine. Ironically, it was produced for use as a chemical warfare agent in World War I, and only later found to be an inexpensive antibacterial, when added to water. Today, chlorinated drinking water is taken for granted throughout the developed world, and has been a leading tool in the fight against waterborne diseases.

6.5.1 Uses of chlorine

The United States Geological Survey does not list chlorine and its uses in its Mineral Commodity Summaries 2012, because salt is listed and tracked. However, chlorine is used in a large variety of applications, including the production of vinyl chloride and thus polyvinylchloride (PVC), as well as a large number of organo-chlorine molecules that are themselves used as intermediates to other end products.

Chlorine continues to be used to purify drinking water, and to be used in the production of bleach.

The chlorination of water for drinking is represented by the following reaction:

$$H_2O(l) + Cl_2(g) \longrightarrow HOCl(aq) + HCl(aq)$$

This is often thought of as an equilibrium, in which OCl^- is favored under basic conditions, whereas $HOCl$ and Cl_2 are favored under more acidic conditions.

6.6 Hydrochloric acid

Hydrochloric acid is usually sold as a 25–38 % by weight solution in water. What is called hydrogen chloride is actually the pure material, often made by the direct addition of elemental hydrogen and elemental chlorine.

6.6.1 Production of hydrochloric acid

The chlorine gas produced as one of the three end chemicals in the Chlor-Alkali Process must be kept separate from the hydrogen that is a co-product, as their combination results in hydrogen chloride. However, the controlled combination of these two under ultraviolet light, along with their absorption into deionized water, gives very pure hydrochloric acid.

The reaction chemistry can be represented simply, as:

$$Cl_2(g) + H_2(g) \longrightarrow 2HCl(g)$$

This method of manufacture long ago replaced the addition of sulfuric acid to sodium chloride, which does indeed produce hydrochloric acid and sodium sulfate.

The production of hydrochloric acid also occurs as a by-product of the chlorination of several organics. The displacement of a hydrogen for a chlorine at a carbon atom results in the chlorinated organic, but also in HCl as the by-product. This is then captured and dissolved in water, resulting in hydrochloric acid. The generic reaction chemistry for this is:

$$R{-}H + Cl_2 \longrightarrow R{-}Cl + HCl$$

6.6.2 Uses of hydrochloric acid

The pickling of steel is a major use of hydrochloric acid, as is the chlorination of organic materials. When steel is pickled, iron oxides are removed from the surface, to produce a clean, reduced surface that can be further treated, painted, or coated. The reason for chlorinating organic materials is that the carbon – chlorine bond can be much more labile than the carbon – hydrogen bond, making organo-chlorides useful synthons for fine chemical production.

The reaction chemistry for these two major uses is as follows:

1. A representative example for the pickling of steel:

$$Fe + Fe_2O_3 + 6HCl \longrightarrow 3FeCl_2 + 3H_2O$$

2. An example for the chlorination of an organic, the production of ethylene chloride:

$$4HCl + 2CH_2 = CH_2 + O_2 \longrightarrow 2ClCH_2CH_2Cl + 2H_2O$$

6.7 Titanium dioxide

This may seem to be an odd chemical to connect with a chapter discussing sodium chloride, but the chlorine that is produced in the Chlor-Alkali Process finds a very specific place in the production of titanium dioxide.

This is a white material used as a pigment in a variety of ways, in both industrial processes, and in many consumer, end-use products such as paints and food.

6.7.1 Production of titanium dioxide

Production of titanium dioxide in usable grades is via two methods: the chloride process, and the ilmenite process.

The chloride process begins with rutile ore, and can be represented by a simple series of reactions, shown below:

$$3TiO_2 \text{ (crude)} + 6Cl_2 + 4C \longrightarrow 3TiCl_4(l) + 2CO_2 + 2CO \qquad 900\,°C$$
$$TiCl_4 + O_2 \longrightarrow TiO_2 + 2Cl_2 \qquad 1,500–2,000\,K$$

Note the titanium chloride is produced as a liquid. This makes it convenient for distillation to be used as a means of purifying it. Note also that the chlorine is regenerated in the process. This cuts down on the overall need for a continuous feedstock of new elemental chlorine.

Additionally, in this process, aluminum chloride can be added as a rutile promoter (the crude TiO_2).

The ilmenite process chemistry is somewhat more complex, but this is because the initial ore has a lower concentration of titanium dioxide. It is represented as follows:

$$2H_2SO_4 + FeO·TiO_2 \longrightarrow FeSO_4 + TiOSO_4 + 2H_2O$$
$$2H_2O + TiOSO_4 \longrightarrow TiO_2·H_2O + H_2SO_4$$
$$TiO_2·H_2O \longrightarrow TiO_2 + H_2O$$

As is seen above, the initial ore must have the iron and titanium separated via chemical means, so that the titanium can be isolated as titanium dioxide. Note that sulfuric acid, the subject of Chapter 2, is a commodity chemical that is used in this process, while never appearing as a product.

The final step requires heat to drive off the associated water.

6.7.2 Uses of titanium dioxide

Titanium dioxide has a variety of uses, which are shown graphically in Figure 6.3 [1].

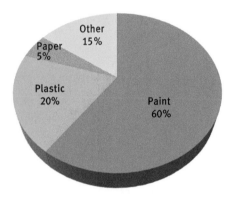

Figure 6.3: Uses of Titanium Dioxide [4].

The "other" includes: "catalysts, ceramics, coated textiles, floor coverings, ink, roofing granules" [4].

The most common use of titanium dioxide of which the general public is aware is as a white pigment. While numerous other materials have been tried as white pigment, titanium dioxide, or titanium white, remains the pigment of choice because of its brightness. It finds uses in paints and a variety of food products, including gum and candies.

6.8 Recycling

Since the production of industrial caustic is an extremely large scale process, almost as big as the production of sulfuric acid, the cost is low enough that NaOH is not regularly recycled. In addition, most sodium hydroxide is used in some further reaction chemistry, which means there is no end material that needs to be recycled.

Chlorine, as we have just seen, also has a variety of uses in which it is reacted, and thus does not exist as elemental chlorine when the reaction is complete (such as the production of chloro-organic chemicals). However, many industries have found it economically effective to capture chlorine that is produced in a process, for further use or recycled use. In the developed world, regulatory agencies monitor the release of chlorine to the atmosphere, and governments may levy fines for its accidental release. Thus, the recovery and re-use of chlorine is often driven by economic factors.

Hydrogen, the third product from the electrolysis of brine, is always either used to generate heat, or consumed in some further reaction. In heat generation with hydrogen, the end product is usually steam or water, and thus there are no recycling programs for the material.

Bibliography

[1] CEFIC. Website. (Accessed 25 October 2022, at www.cefic.org/).
[2] The Chlorine Institute. Website. (Accessed 25 October 2022, at https://www.chlorineinstitute.org/).
[3] Eurochlor. Website. (Accessed 25 October 2022, at https://www.eurochlor.org/).
[4] Mineral Commodity Summaries 2022, U. S. Department of the Interior, U. S. Geologic Survey, Reston, VA, USA, https://pubs.er.usgs.gov/publication/mcs2022.
[5] Salt Association. Website. (Accessed 25 October 2022, at https://saltassociation.co.uk).
[6] The World Chlorine Council. Website. (Accessed 25 October 2022, at www.worldchlorine.org/).

7 Further inorganics

7.1 Carbon black

Carbon black is known by several common trade names, including: acetylene black, channel black, colloidal black, furnace black, and thermal black, depending upon its source material. It is routinely produced through incomplete combustion of the heavier fractions of petroleum, such as coal tar or the heavy oils. It can however be produced through the lean combustion of organic materials such as vegetable oils.

Because carbon black is produced from a variety of hydrocarbon sources, it can be classed as an organic material. However, because it has no covalent bonding to any other elements, it can just as fairly be classified as an inorganic material. This wide variety of source materials makes a representative reaction chemistry for its production difficult to simplify. One example however, might be:

$$H - (CH_2)_n - H \longrightarrow H_2 + C_{(s)}$$

This ignores the traces of carbon monoxide that can be found in the gas stream after the reaction is complete.

7.2 Producers

Carbon black is produced and marketed on a large enough scale that a manufacturers organization exists related to it, the International Carbon Black Association (ICBA) [8]. Member companies of the ICBA are, in alphabetical order: Alexandria Carbon Company, Cabot Corporation, Cancarb Limited, Columbian Chemicals, Continental Carbon, Evonik, Sid Richardson, and Timcal Graphite and Carbon.

Alexandria Carbon Company, under the larger firm Aditya Birla, has been able to reach production levels of 285,000 tons per year at a single plant in Egypt.

Cabot Corporation claims on its web site, "As the world's largest manufacturer, Cabot produces nearly 2 million metric tons of carbon black annually and sells it under the following trade names: BLACK PEARLS, ELFTEX, VULCAN, MOGUL, MONARCH, EMPEROR and REGAL." Clearly, the market for carbon black is a robust and world-wide one.

Cancarb Limited, headquartered in Alberta, Canada, is also capable of producing 45,000 metric tons per year, which they label "thermal carbon black," and claim to be the largest operation of its kind.

Columbian Chemicals, based in Marietta, Georgia, USA, is also part of the Aditya Birla Group, and markets several grades of carbon black under a variety of trademarked names, including: Raven®, Conductex®, Copeblack®, Furnex®, Statex®, and Ultra®.

Continental Carbon operates plants in the United States, Taiwan, China, and India. As might be expected, it has also begun to expand into carbon nanotubes, although sales of the latter are not yet nearly as large as the sales of carbon black [1, 2, 3, 4, 5, 6].

https://doi.org/10.1515/9783110671094-007

Evonik, a long-time carbon black producer, was sold on April 16th, 2011. According to the press release: "Evonik Industries signed an agreement with affiliates of Rhône Capital LLC to sell its carbon black business, which achieved sales revenues of some €1.2 billion in 2010. The transaction is valued in excess of €900 million including the assumption of certain obligations." [7] Evonik remained listed as a member of the ICBA for some time.

Sid Richardson, operating out of Texas, USA, produces more than 30 grades of carbon black, depending on market and customer needs [13].

Imerys Graphite and Carbon has numerous sites in North America, Europe, and eastern Asia, and produces several other carbon products in addition to carbon black [15].

According to the International Carbon Black Association (the ICBA): "Two carbon black manufacturing processes (furnace black and thermal black) produce nearly all of the world's carbon blacks, with the furnace black process being the most common." [8]

7.3 Production methods

The furnace black process functions with the heavier aromatic oils and tars as its principle feed materials. This is economically more favorable than conversion of this fraction into octane. The heavy oil is vaporized and combined with a hot gas stream, usually created using a natural gas source, under carefully controlled temperature and pressure conditions. The carbon particles that form in these conditions must be cooled, and then collected. The collection process also filters out other products. These are usually gases such as hydrogen, which must be stripped from the feedstock, or carbon monoxide, which results from incomplete combustion, despite the temperature and controlled conditions of the reaction chamber. These secondary products are usually recycled to a use such as power or heat generation.

In a step-wise fashion, the process can be delineated as follows:
1. Feedstock oils are passed through a chamber heated by natural gas to 3,500 °C.
2. This chamber generates a vapor stream of the feedstock oil at high temperature. Formation of carbon black particles begins here.
3. The vapor passes from the chamber to the bag filter house, where several thousand filter bags seize and separate the carbon black from the gas stream.
4. The remaining material from the vapor stream is further filtered, so that it can be released into the atmosphere.
5. The carbon black particles are shaken from the filters in the bag house, dropped into a collection pipe and pelletized.
6. The carbon black particles are combined with a water mist to create larger carbon black particles.
7. The larger particles are then dried to reduce moisture content, and packaged for sale.

Slight variations of the process result in varying sizes of carbon black pellets.

The other major process for carbon black production, the thermal black process, Cancarb Limited describes very succinctly as:

"The process of producing thermal carbon black involves decomposing natural gas in the absence of oxygen. Each of Cancarb's five production units includes two reactors. One reactor is preheated to about 1,300 °C, and natural gas is injected into it and is decomposed into carbon and hydrogen. The carbon/hydrogen mixture is cooled with water and the carbon is separated from the hydrogen in large bag filters. The hydrogen by-product is then used as fuel to preheat the second reactor of the unit. As the process of producing the carbon consumes energy, the first reactor will eventually cool to a point where the reaction becomes inefficient. This is when the reactors trade roles and the second one becomes the producing reactor while the first one is reheated." [4]

Note the use of the by-product hydrogen to make the entire carbon black production more efficient in terms both of material and costs.

7.4 Uses

As with many large commodity chemicals, carbon black ultimately has numerous end uses, although one broad category, tires, overwhelms all the others combined. Carbon black is not used simply to give automobile and truck tires their black color, but to reinforce and strengthen them.

Broadly, the uses of carbon black can be broken down as follows, below, and also in Figure 7.1:
1. Reinforcing agent in vehicle tires, 90 %.
2. Plastics, printer inks, and toner, as pigments, 9 %.
3. Electrostatic discharge compounds.
4. High performance coatings.

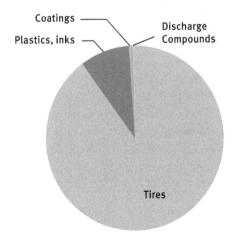

Figure 7.1: Uses of Carbon Black.

7.5 Recycling

We have already mentioned the re-use of byproduct hydrogen for heating the carbon black processes. Since virtually all carbon black is consumed in some further process, resulting in some end use product, there are no recycling programs for the material. The recycling of tires is addressed in Chapter 21, Rubber.

7.6 Potash

The term 'potash' is used for several potassium-containing minerals, all of which are used almost exclusively for fertilizer, and all of them water soluble to a great extent [10]. In this context, potassium chloride is sometimes referred to as muriate of potash, or MOP, and potassium sulfate is sometimes referred to as sulfate of potash, or SOP. The Potash Development Association [11, 12], a British based trade organization, is involved with fertilizer use and applications in Great Britain, and indicate that the organization is supported by fertilizer manufacturers internationally who supply the British fertilizer market [12].

7.6.1 Producers

Cleveland Potash is a major producer of potash for the fertilizer industry [5], with mining capability in Great Britain of over 1 million tons of potash for fertilizer per year.

Tessenderlo Group produces a variety of chemicals, but includes four trademarked names of potash in their sales profile [14].

Potash Corp., headquartered in western Canada, is another large fertilizer company, which claims to be the largest worldwide, producing not only potash, but phosphate and nitrogen-containing chemicals for fertilizers [11, 12, 13, 14]. PotashCorp claims to be producing 20 % of the world capacity. The firm uses conventional underground mining to produce much of the potash that is mined in Canada.

Western Potash uses solution mining for its operations, as seen in Figure 7.2, also located in western Canada [16]. The process involves two wells through which hot water is injected, and brine is then extracted. The process works well because of the solubility of the potassium ores in water [17]. Additionally, solution mining operations tend to be less expensive than traditional mining operations, because less equipment has to be placed underground, and there are less human safety elements with which to deal. Additionally, less waste is generated in solution mining operations, because waste material can be reintroduced to the mine as a slurry. Thus, there is minimal need for above ground disposal of excess by-product or waste materials.

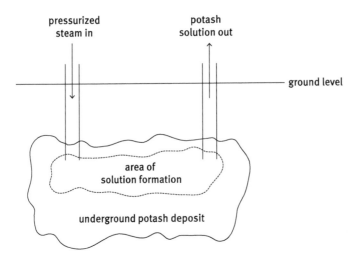

Figure 7.2: Solution mine.

Within the United States, potash is produced in several states, including Michigan, Utah, and New Mexico [10]. Ores can either be processed and crystallized by surface evaporation, or can be extracted through "deep-well solution mining" [10].

7.7 Production by country

World production and reserves are shown in Table 7.1, below, and have been tabulated by the United States Geologic Survey [10]. Countries not on the list are assumed to be small producers or net importers of the material.

Data from the USGS for potash is in metric tons of K_2O equivalent [10].

7.8 Recycling

Because potash is used almost exclusively as a fertilizer, and thus all of the material is sent for end use to farms worldwide, there are no recycling programs in place for this material.

7.9 Sodium tri-poly-phosphate

Sodium tri-poly-phosphate, often abbreviated STPP, is a major component of multiple types of detergents. It is also used as a food preservative, and has several other uses that are not as large as these two. As a detergent component, the tri-poly-phosphate carries a high enough negative charge that it can chelate dications, the cause of hard water,

Table 7.1: Potash Production, Worldwide.

	Mine production 2020	Mine Production 2021	Reserves
U.S.	460	480	970,000
Belarus	7,400	8,000	3,300,000
Brazil	254	210	10,000
Canada	13,800	14,000	4,500,000
Chile	900	900	
China	6,000	6,000	
Germany	2,200	2,300	
Israel	2,280	2,300	
Jordan	1,590	1,600	
Laos	270	300	500,000
Russia	8,110	9,000	
Spain	420	400	
Others	360	370	1,500,000
World total	44,000	46,000	11,000,000

Blanks indicate data was not provided.

and thus remove them from water. This allows the detergent, usually a long, sulfonate chain, to function without any deactivation. In the recent past, over 4.5 million tons are produced annually.

The use of detergents has become so prevalent in modern society that there are several trade organizations dedicated to detergents, such as the American Cleaning Institute, which predictably emphasizes the higher standard of living people enjoy because of the uses of detergents [2], the UK Cleaning Products Industry Association [18], and the Japan Soap and Detergent Association [9], which have similar focuses.

7.10 Producers

Worldwide, there are a large numbers of producers of sodium tripolyphosphate. Some of the larger companies include:
1. Ortho Chemicals Australia Pty, Ltd, based in Victoria.
2. Thatcher Company, based in Salt Lake City, Utah.
3. Hubbard Hall, based in Connecticut.
4. Biddle Sawyer Corporation, with offices in New York City.
5. Prayon Deutschland, GmbH, based in Dortmund.
6. Tianjin Ronghuiyuanyang International Trade Co., Ltd, based in Tianjin City.

The market is large enough, and there is enough demand for the chemical by industry that there are producers in most countries of the developed world.

7.11 Production

STPP can be produced from the reaction of phosphoric acid and sodium carbonate, the production of the latter having been covered in Chapters 5 and 6.

$$H_3PO_4 + Na_2CO_3 \longrightarrow Na_5P_3O_{10}$$

7.12 Uses

There is a wide number of different intermediate and end uses for STPP. In general however, uses for STPP fall into the following broad categories:
1. Detergents, by far the largest share.
2. Food-grade, for moisture retention.
3. Water softeners.
4. Clay processing.
5. Textile processing.
6. The paper industry, pulping.
7. Rubber manufacture.
8. Paint formulations.
9. Ore flotation.

7.13 Recycling

Like the other two materials discussed in this chapter, STPP is produced for use in a variety of consumer products, which means it is never present in great enough or concentrated enough quantity after use that there is any economic incentive to recycle it.

Pollution from the use of excessive amounts of detergents has been known to cause extensive environmental damage, and the development of methods to eliminate the discharge of such materials into open waterways remains an important goal for wastewater treatment facilities to achieve.

Bibliography

[1]   Aditya Birla Carbon Black. Website. (Accessed 27 October 2022, at www.birlacarbon.com/).
[2]   American Cleaning Institute. Website. (Accessed 27 October 2022, at www.cleaninginstitute.org/).
[3]   Cabot Corporation Specialty Carbon Blacks. Website. (Accessed 27 October 2022, at https://www.cabotcorp.com/solutions/products-plus/specialty-carbons/for-plastics#:~:text=Specialty%20carbon%20blacks%20are%20produced,their%20particle%20size%20and%20structure).
[4]   Cancarb Limited. Website. (Accessed 27 October 2022, at https://cancarb.com/products/).
[5]   Cleveland Potash. Website. (Accessed 27 October 2022, at https://www.iclfertilizers.com/browse).
[6]   Continental Carbon. Website. (Accessed 27 October 2022, at www.continentalcarbon.com/).

[7]   Evonik sale. Website. (Accessed 14 October 2012, at corporate.evonik.com/en/media/press_releases/
       pages/newsdetails.aspx?newsid=19107).
[8]   International Carbon Black Association. Website. (Accessed 26 November 2022, at carbon-black.org/).
[9]   Japan Soap and Detergent Association. Website. (Accessed 26 November 2022, at jsda.org/w/e_engls/
       index.html).
[10]   Mineral Commodity Summaries 2022, United States Geological Survey. Website. (Accessed 26
       November 2022, as: pubs.er.usgs.gov/publication/mcs2022).
[11]   Nutrien. Website. (Accessed 26 November 2022, at: www.nutrien.com).
[12]   Potash Development Association. Website. (Accessed 26 November 2022, at www.pda.org.uk/what-is-
       potash.html).
[13]   Sid Richardson. Website. (Accessed 26 November 2022, at crunchbase.com/organization/sid-
       richardson-carbon—energy.co).
[14]   Tessenderlo Group. Website. (Accessed 26 November 2022, at www.tessenderlo.com).
[15]   Imerys Graphite and Carbon. Website. (Accessed 26 November 2022, at www.imerys.com).
[16]   Western Potash. Website. (Accessed 26 November 2022, at www.westernpotash.com/).
[17]   Western Potash Solution Mining. Website. (Accessed 26 November 2022, at www.westernpotash.com/
       about-potash/solution-mining).
[18]   UK Cleaning Products Industry Association. Website. (Accessed 26 November 2022, at www.ukcpi.
       org/).

8 Water

8.1 Introduction, sources

This chapter is devoted to what might be the single most useful, common chemical on earth (one might argue that air or silicate minerals are as important, but many would insist on water). While more than 95 % of the world's water is salt water, almost all water used industrially and residentially starts as fresh water, and may have to undergo purification before it is considered usable for a specific application. There are however, different levels of purity of fresh water, all of which have useful applications in industry. Water used for cleaning, for example, is far less pure than water required in any step in the fabrication of computer chips or other end goods manufactured to extremely high standards.

8.2 Purification techniques

Because water is so ubiquitous and so vital, a variety of ways have been developed by which it can be purified, and several organizations exist that monitor water quality [3, 2, 1, 9, 4]. The methods we detail, below, are not all inclusive, but do represent all the major means by which water is desalinated or re-purified after use.

8.2.1 Desalination

There are a variety of desalination techniques, all of which result in clean, drinkable water. The choice for which to use is often dictated by both location – and thus what the intake water has as contaminants – and cost [5, 6, 7]. Desalination apparatus can be as small as a hand-held item (often included in the safety equipment of ocean-going recreational small boaters), or as large as the apparatus of any major industry. One of the largest desalination plants is in the Mediterranean Sea, off the coast of Haifa, Israel, although other large plants are used to produce drinking water in Singapore, Aruba, and several other locations.

Multi-stage flash distillation accounts for more than 75 % of desalination apparatuses. This is not a chemical separation of salts from water; rather, it is separation based on a series of heat exchangers causing small amounts of water to turn to steam (to "flash") at each chamber of the apparatus. The major long-term expense in this process is the energy needed to generate steam. The basic scheme is shown in Figure 8.1.

Reverse osmosis is also used to purify water, involving a semi-permeable membrane through which water is forced, but through which particles are not able to pass. The osmotic pressure that will eventually dilute brackish water on one side of a membrane by allowing mixing with clean water from the opposing side of the membrane is overcome

https://doi.org/10.1515/9783110671094-008

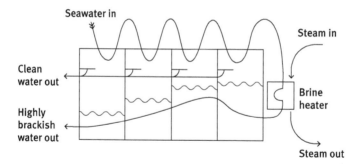

Figure 8.1: Water Desalination Process.

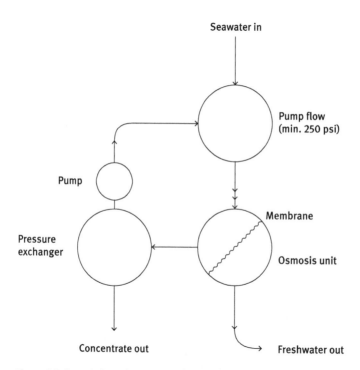

Figure 8.2: Reverse Osmosis Water Purification.

by an application of force, thus concentrating the brackish side, and yielding clean water on the opposing side, as seen in Figure 8.2.

Large reverse osmosis plants are used in areas of the world where a population lives in or near a desert, but where the ocean is also close. Much like other forms of desalination, the cost of energy to force the process is a major expense. Much smaller reverse osmosis water purification units (called ROWPU) are used by some militaries in the field, so that they are able to accomplish their missions without having to bring all their potable water with them by train, truck, or plane.

There are a variety of other water desalination processes, including:
- membrane distillation
- electrodialysis reversal
- nano-filtration
- ion exchange
- solar desalination
- freezing desalination

All of these processes are effective in removing salt from water, but all have various costs associated with them. In almost all cases, the search for lower energy costs is a continuing one that can ultimately improve each process.

8.2.2 Sewage treatment

Sewage treatment can be considered both a form of cleaning water, and a means by which materials dissolved in water are prevented from entering the environment in an undesired manner. When water is discharged and other materials are captured, sewage treatment is considered to be a final process for the water. When the water is re-captured and re-used, sewage treatment is also considered a form of recycling for the water.

There have been several dramatic news reports on American television in the past decades in which an on-the-street reporter offers passers-by glasses of clean, fresh water, then informs them that the water came from a sewage recycling system. The idea is to raise awareness that treated water from a sewage plant is entirely healthy and palatable to consume, and to try to overcome people's distaste and disgust when they connect drinkable water with sewage.

Treatment of sewage can be broken into different techniques and methods, which are generally based on the water-based material that comes into the treatment plant, and the availability of different treatment capabilities in different areas (such as wet areas, temperate areas, or dry, desert environments). Essentially all sewage treatment plants do however have these steps in common:
- Pre-treatment. This involves the removal of macroscopic contaminants. Since sewage can include storm water run-off, this step includes the removal of yard waste, such as grass, seeds, and leaves. In urban areas this can also include the removal of small but macroscopic trash.
- Primary treatment. This usually involves some form of settling, so that heavier-than-water materials sink to the bottom of the treatment container, and materials such as oils or fats can rise to the top for separation and removal.
- Fat and oil removal. (if necessary) Fats, oils, and greases can be removed with manual skimmers, or by injecting air into the water batch to induce a froth which can then be separated.

- Secondary treatment. This routinely includes some form of microbial digestion of biological material that is either dissolved or suspended in the water. Often, after a suitable time for digestion, the tanks are drained, leaving any algae or other material that has grown during this step to dry, then be collected and removed.
- Sludge digestion. The water that is produced at this point is clean enough for return to the environment (although it usually requires further steps such as chlorination to make it fit for drinking), leaving behind what is referred to as sludge. If this is not dried and disposed of in some manner, it can be used in some application, such as a fertilizer, if the sludge is rich enough in plant nutrients.

It can be seen from this list that the purification of water from sewage is less of a series of chemical reactions than it is a series of steps based on physical differences and biological reactions. For many years, the purpose of sewage treatment was to prevent the spread of disease. The production of clean water was either a by-product or an afterthought of the process.

8.3 Uses, residential

The developed world is marked by water distribution systems from municipal points to personal residences. There are very few families in what can be called The First World that still have to bring water into a home that has no internal plumbing. In most areas, fresh water is used as the source, cleaned of contaminants, and pumped to distribution points, then to residences. Areas lacking adequate fresh water sources can use salt water sources, but obviously, desalination is required before distribution from the source to any distribution points. Israel currently has one of the world's largest reverse osmosis desalination plants, off the coast of Haifa, which is used to provide fresh drinking water for personal use, although Singapore also uses water desalination plants extensively [7].

8.4 Uses, industrial

We have already seen the largest industrial use of water as a reactant, in the production of sodium hydroxide and chlorine, in Chapter 6. There are numerous other processes however that require water, although not as a reactant. Oil drilling and processing require significant amounts, as does agriculture.

8.5 High purity water, uses

High purity water is used in the production of various electronics and components of electronics, and must contain essentially no soluble ions. While the production of ultra-high purity water is not yet one of the largest chemical manufacturing processes, the need for it continues to increase, as more applications for precision manufacturing come on line.

In the production of high purity water, cation exchange membranes are thus used to remove cations and replace them with hydronium ions. Similarly, anion exchange membranes can be used to replace anions in solution with hydroxide ions. The end result is water free of any cation or anion contaminants. The process usually requires multiple steps in which the water passes through cation and anion exchange membranes in tandem.

Electrodeionization is another technique by which ions can be remove from water, in this case by having water pass between two oppositely charged electrodes. The resultant purity is comparable to water purified by ion exchange.

Finally, the direct combination of pure elemental hydrogen and pure elemental oxygen can be used to generate high purity steam, in which the resulting water is even free from any traces of dissolved oxygen. This is suitable for the production of small amounts of extremely high purity water, but must be monitored carefully because of the risk of explosion.

8.6 Recycling

In general, while the technology exists for large scale desalination of water, and treatment of waste water, in terms of energy and cost, simple conservation of existing clean, potable water remains roughly 100 times more cost effective than either technique. Of the two, waste water treatment remains less expensive than desalinization (but it is difficult to quantify the cost savings, since the degree of treatment needed for polluted, non-saline water differs by location) [8].

Water treatment plants, and thus the recycling of water, is an enormous industry in and of itself, one that uses various chemicals, but one that also uses remarkably intelligent low-tech solutions to return water to a state that is either clean enough to re-use, or clean enough to discharge back into a natural environment. Yet the very success and reliability of water treatment plants in the developed world has led to a complacency about the availability of clean, potable water. Additionally, the use of a wide variety of materials that were not in existence when many water and sewage treatment plants were built has resulted in problems with effluent streams, such as unreacted pharmaceuticals in water.

Waste water and sewage treatment plants for many municipalities treat incoming material in three broad stages. The primary stage removes macroscopic, solid materials of all kinds. The second stage often allows or permits some form of algal growth, which is then filtered through sand, allowing the algal bloom to dry and die. This is then collected and disposed of. The tertiary stage involves the injection of an antibacterial, such as chlorine, into the water so that it may be re-used. This stage is not necessarily performed if the water is to be discharged back into the environment.

8.7 Further

The shortcomings of existing water treatment plants, specifically for the removal of waste or unused pharmaceuticals, is discussed further in Chapter 19. Future treatment plants will have to be designed to enable the removal of persistent, water stable organic molecules in order to purify water to a point where it is useful for human consumption and industrial use, or for discharge into natural waterways. The current inability to do so has produced situations in which pharmaceuticals are found in river and lake water, or in drinking water. The long-term effects of this are not yet known.

Bibliography

[1] American Water Resources Association. Website. (Accessed 27 November 2022, at www.awra.org/).

[2] American Water Works Association. Website. (Accessed 27 November 2022, at www.awwa.org/index. cfm?showLogin=N).

[3] Aqua Europe. Website. (Accessed 27 November 2022, at www.aqua-europa.eu/).

[4] British Water. Web site. (Accessed 27 November 2022, at www.britishwater.co.uk/).

[5] Desalination: A National Perspective. National Academies Press, 2008. (Accessed 27 November 2022, at: nap.nationalacademies.org/catalog/12184/desalination-a-national-perspective).

[6] International Desalination Association. Website. (Accessed 27 November 2022, at www.idadesal.org/).

[7] Singapore Water Association. Website. (Accessed 27 November 2022, at www.swa.org.sg).

[8] Snyder, S. (Organizer). "The Business of Water: Problems, Solutions, an Opportunities for the Chemical Enterprise". Council for Chemical Research, 30th Anniversary Annual Meeting.

[9] Water Quality Association. Website. (Accessed 27 November 2022, at www.wqa.org/).

9 Simple organics from petroleum

This chapter will describe the fractionation processes used on crude oil, and lead to the further chapters on petroleum-based materials.

9.1 Introduction and history

In various parts of the world, oil has seeped to the Earth's surface naturally throughout history. Some older accounts refer to crude oil that has come to the surface as naphtha, although the word acquired a different meaning when the drilling and refining of crude oil became an industry. For well over a century, people have realized that crude oil can be separated into different fractions and components. While a large fraction of crude oil that is distilled today is refined into motor gasoline (what is called the octane fraction), the rough equivalent of modern gasoline had been isolated in the mid-1800s, and had a variety of uses prior to becoming the major motor fuel. The fact that gasoline was already a commercially available commodity as early automobile engines were being developed means that this is one of only a few cases in history where some mechanical device was developed around the existing fuel for it (railroad engines or steamship engines and coal is another).

As the fractionation of crude oil became more efficient and more precise, a variety of chemicals have been isolated from it that gave rise to what is now an enormous worldwide plastics industry. Polyethylene, polypropylene, polystyrene, nylon, and numerous other synthetic polymers are made from these isolated materials. This will be discussed in detail in Chapter 15. For the most part, there are only seven materials isolated from crude oil from which virtually all others are produced. They are: methane, ethylene and propylene, the C4 fraction, and the three aromatics: benzene, toluene, and xylene. The largest derivatives made from each are shown in Table 9.1.

9.2 Sources, geographically

It is difficult to believe an educated person today could have no knowledge of where in the world oil is located, because the political and economic situation in today's world is often affected by the presence of oil deposits. The Organization of the Petroleum Exporting Countries, OPEC, currently has the following nations as members, listed alphabetically: Algeria, Angola, Congo, Equatorial Guinea, Gabon, Iran, Iraq, Kuwait, Libya, Nigeria, Saudi Arabia, United Arab Emirates, Venezuela [2]. Perhaps obviously, different nations have different outputs on a daily and annual basis, and each country relies on its oil exports to a different extent when output is measured as part of the gross national product. Of these nations, Iran, Iraq, Kuwait, Saudi Arabia, and Venezuela are the founding member countries, based on the OPEC agreement signed in Baghdad in September

https://doi.org/10.1515/9783110671094-009

Table 9.1: Chemicals Made From Hydrocarbon Starting Materials.

Hydrocarbon	Derivatives
Methane	Acetic acid
	Dimethyl terephthalate
	Formaldehyde
	Methanol
	Methyl-t-butyl ether (MTBE)
	Vinyl acetate
Ethylene	Acetic acid
	Ethylene dichloride
	Ethylene glycol
	Ethylene oxide
	Ethylbenzene
	Styrene
	Vinyl acetate
	Vinyl chloride
Propylene	Acetone
	Acrylonitrile
	Cumene
	Isopropanol
	Phenol
	Propylene oxide
Butyl fraction	Acetic acid
	Butadiene
	Methyl-t-butyl ether (MTBE)
	Vinyl acetate
Toluene	Benzene
Benzene	Acetone
	Adipic acid
	Caprolactam
	Cumene
	Cyclohexane
	Ethylbenzene
	Phenol
	Styrene
Xylene	Dimethyl terephthalate
	Terephthalic acid
	p-Xylene

of 1960. But other countries have been significant producers of petroleum, and are not necessarily members of OPEC. The United States of America, Great Britain, and Russia are three such.

Oil exploration continues in widely separated areas of the world today, because the need for both the materials that will be used to form plastics and polymers, and the apparently never-ending need for a variety of motor fuels. From northern Canada and the fields of Alaska, to the Gulf of Mexico and Indonesia, exploration continues no matter the climate or whether the known deposit is below land or sea.

9.3 Extractive techniques

Since so much of the world's energy and finished plastic products are currently dependent upon petroleum as a feedstock, we need to consider the different means of extracting petroleum.

The American Petroleum Institute generally divides extractive techniques or zones by source material, into the following categories: onshore, offshore, oil sands, oil shale, and natural gas, and hydraulic fracturing [1].

"Deep water drilling" is a term that appears to get deeper as advances in rig platforms and other, related technologies progress. Currently, drilling can take place in 10,000 feet of water, but the fixed, offshore platform one associates with any offshore drilling is being substituted by an array of novel, floating platforms and systems. Drilling in depths up to 2,000 feet has become common and established, and now uses proven technologies. Different designs and types of drilling systems are discussed below.

9.3.1 Offshore

There are numerous variations in design and type of operations that bring crude oil from undersea deposits to land-based refineries.

- Compliant Towers. The compliant tower design has been developed so that the entire drilling operation can withstand the back-and-forth forces (lateral stress) of ocean waters. The production deck does not differ appreciably from the older, more traditional fixed platforms. The compliant tower design allows drilling in water depths of up to 2,000 feet.
- Fixed Platform. As the name implies, such platforms have large, steel supports driven into the seabed to fix them in place while crude oil is extracted. The supports are then fitted with all the necessary drilling, production, and crew facilities on the platform above. Such platforms are currently used in numerous operations when water depth does not exceed 1,500 feet.
- Floating Production Systems. Again, as the name implies, these systems float on the sea surface, or are partially submerged to increase stability, but do not have fixed steel supports to the seabed. They do require some form of connective chain or cabling to keep the surface unit proximate to the drilling site. Because of their flexibility, floating production systems are currently used for the deepest water drilling sites.
- Floating Production, Storage, and Offloading Systems. This type of system requires a floating tanker storage unit, usually moored to the well site. Its greatest advantage

is that it can be used to store crude from wells that do not produce large enough amounts that a direct pipeline to a distribution facility is warranted.

– Spar Platform. This type of system utilizes one cylinder that rises vertically and supports the above-water deck and operations. Because of the strength of the system, it can be utilized for operations that normally extend to 3,000 feet, and can be used for much deeper sites if necessary.

– Subsea Systems. This term simply refers to wells, either single or multiple, that remove crude from below the seabed to a floating platform system. Generally, subsea systems are used in situations where water depth is extreme.

– Tensions Leg Platforms. (and mini-tension leg) TLPs utilize a floating platform, a base in the seabed, and what are termed "tensioned tendons" that connect the two. The tendons can be clustered together to improve stability of the entire operation. Using this method, drilling and extraction can take place in up to 4,000 feet of water, although this technique is also well suited to much shallower depths. It has been used in North Sea operations, which can be as shallow as 100 meters.

The mini-tension leg platform utilizes the same technology, but tends to be employed for smaller reserves which would otherwise not be economically feasible to recover.

9.3.2 Onshore

As with offshore drilling and recovery of crude oil, there are a variety of different means by which oil is extracted from land-based sites.

– Oil wells. Oil wells are sited after extensive research into the geology of a specific area. When possible, refineries can be located near wells, but this is not required, as the shipping of ocean going tankers proves. One continued disadvantage of having to ship raw material to refineries is that it is not always economically practical to ship the lightest fraction, because it exists in a gas state. When this is the case, the material is burned at the site, thus losing this fraction of the crude for any further, possible use.

– Oil sands. Besides traditional oil wells, oil sands have become zones of increasing interest in the past decade. The material to be extracted is often called bitumen or tar, and is extremely viscous. Thus, energy must be put into the extraction process, as well as various chemicals and a significant quantity of water. Shallow deposits of oil sands are actually strip mined.

Kazakhstan, Russia, and Canada have significant deposits of oil sands, although the Canadian deposits are by far the largest, and thus the most economically profitable, with the ability to produce well above 100 million barrels per day. The Russian and Kazakhstani deposits are located in relatively close proximity to their conventional oil deposits.

- Oil shale. The production of petroleum from oil shale is a matter of coupling a mining operation with refining after the product is mined. Strip mining, open pit mining and underground mining of oil shale are all possible.

  Treatment of the oil shale can be performed underground or above ground. The process involves heating the material to approximately 500 °C in a lean, oxygen-free atmosphere to ensure breakdown of the raw material into gases, a liquid phase that can be refined, and solid, residual material.
- Hydraulic Fracturing. Often called 'fracking,' the process of hydraulic fracturing pushes oil deposits from rock formations through the injection of high pressure water and chemical additives into the ground and rock formation. While fracking has existed for decades, horizontal slickwater fracturing has recently allowed the extraction of shale gas from deposits in the United States that have not been economically viable before. This method allows the fracking fluid to be injected into a rock formation in a predominantly horizontal direction, as opposed to a vertical well configuration.

  Proponents point out the large reserves that can be extracted using this fracking technology, while others have voiced concerns about the possible long term effects and pollution associated with forcing fracking fluids into rock formations, which may themselves be part of the larger water table for an area.

The composition of fracking fluids varies depending upon the geologic formation into which the fluid is injected, and what type of fracturing is required for an operation. Components can include those listed in Table 9.2.

Table 9.2: Possible Fracking Fluid Materials.

Material	Role in Fracking Fluid
Borate Salt	Adjust fluid viscosity
Citric acid	Corrosion inhibition
Ethylene glycol	Anti-scaling agent in piping
Guar gum	Adjust fluid viscosity
Gluteraldehyde	Anti-bacterial
Hydrochloric acid	Aiding and creating fissures
Isopropyl alcohol	Adjust fluid viscosity
Polyacrylamide	Reduces friction in flow

Fracking has proven itself to be a useful method by which oil deposits can be profitably extracted from rock formations that have thus far been unrecoverable. But long-term studies of the effects of fracking and fracking residues in rock formations have not yet been undertaken.

9.4 Refining and distillation

The refining and distillation of crude oil is a very mature industry that stretches back almost 150 years. The broad steps by which the thousands of compounds in crude oil are separated can be broken into the following broad steps:

9.4.1 Simple de-salting

This step, usually accomplished at 60–90 °C, simply removes the suspended materials in crude oil such as salts and clays.

9.4.2 Distillation

This step can be accomplished at ambient pressure, and temperatures elevate to no higher than 400 °C. The aim of the step is to begin the separation of the thousands of compounds in crude oil into fractions that can be handled more easily, based on their boiling points.

9.4.3 Hydrotreating or hydroprocessing

This step is performed at elevated pressure (usually 200–300 psi) and temperatures of 350–400 °C. At this stage, the breaking of heavy hydrocarbons to light begins. The addition of hydrogen to the mix is necessary at this step.

9.4.4 Cracking or hydrocracking

This step is an enhancement of hydrotreating, and involves longer contact times than the former. It too involves the production of lower molecular weight hydrocarbons from higher. Both of these steps are designed in large part to enhance the amount of motor fuel that can be derived from the raw material.

9.4.5 Coking

Sometimes called destructive distillation, this step is a form of severe thermal cracking, usually effected at 450 °C. Once again, the aim is to increase the amount of motor fuel that can be extracted. This step also tends to increase the sulfur content of the material. But sulfur content recovery is also strictly monitored and controlled.

9.4.6 Visbreaking

This step, performed on heavy oils at approximately 480 °C, is a further means of breaking down higher molecular weight hydrocarbons.

9.4.7 Steamcracking

This step is often performed at an adjacent facility or refinery, and is designed to produce olefins (alkenes). The step requires 800–850 °C temperatures, and can use feeds as small in molecular weight as ethane, and as large as the heavier oils, such as vacuum gas oil.

9.4.8 Catalytic reformers

This step utilizes naphtha as the feedstock, and runs at 430–500 °C, and pressures of 150–900 psi. Hydrogen is the by-product of lower molecular weight hydrocarbon production and olefin production at this step.

9.4.9 Alkylation

This step reacts lower alkene molecules with paraffin to result in highly branched smaller alkanes. This in turn becomes the prime motor fuel components with the highest octane number, which will be discussed further in Chapter 13.

9.4.10 Removal of the C1 or natural gas fraction

Natural gas can be as high as 97 % methane, and thus removal of the gas fraction and separation into methane and other light gases is an important step. The primary uses for methane are as fuel and as a hydrogen source, which was discussed in Chapter 7.

9.4.11 Sulfur recovery

Sulfur must be removed from crude, in order to minimize the release of sulfur oxides as pollutants. We have seen in Chapter 2 that hydrogen sulfide can have oxygen gas added to it to initiate the production of sulfuric acid. Sulfur in crude oil that is chemically bonded to carbon is first reduced to sulfur or hydrogen sulfide, then converted into sulfur oxides for production of sulfuric acid, if the sulfur content is high enough.

9.5 Pollution

Volumes have been written and continue to be written about how petroleum products, and their combustion, pollute the Earth, its air, land, and waters. Whether it is in the refining and production process, or in the form of carbon dioxide emissions as refined petroleum products are combusted, one can make the claim that almost every aspect of petroleum use involves some form of pollution. To counter-balance such statements, it is fair to note that the quality of life enjoyed by the developed world today is due in large part to petroleum refining and end product use. Choosing just one example, medical procedures would be nowhere near as safe, hygienic, and inexpensive as they are today if plastics made from crude oil source materials were not utilized in almost every procedure. We will delve into the specifics of pollution and recycling more in the following chapters, when we discuss specific fractions of the refining process.

9.6 Recycling and by-product uses

We will discuss the recyclability of plastics in Chapter 15, Polymers. If carbon dioxide is considered a by-product in petroleum use, as opposed to an end product, it can be stated that while there is currently no industrial level re-use of the material, that numerous researchers are working on ways to either capture CO_2, or to in some way utilize it to produce any other chemical from it. It is also fair to say that if a single process can be found that re-converts CO_2 into some more reduced form of carbon – especially one that can be used in some fashion to make a chemical product or generate energy – the world will be forever changed.

Bibliography

[1] American Petroleum Institute. Website. (Accessed 29 November 2022, at www.api.org/).
[2] Organization of Petroleum Exporting Countries. Website. (Accessed 29 November 2022, at www.opec. org).

10 The C1 fraction

10.1 Methane

Methane is the lightest molecular weight material of the petroleum-based hydrocarbons, and is often found in such quantities that its extraction is from what are called natural gas wells, as opposed to oil wells.

It has proven the most difficult of the hydrocarbons to oxidized partially, although we have seen that hydrogen stripping provides one of the raw materials for industrial-scale production of ammonia. This is because the energy required to break the C-H bond in methane is higher than the energy required to break the C-H bond in any oxidized material derived from methane, such as methanol, formaldehyde, or formic acid. Thus, initiating the oxidation of methane means the oxidation products will most likely be carbon dioxide, carbon monoxide, or carbon black.

Methane is routinely used as a source of fuel, with the end-use of home cooking being arguably the most familiar application. Steam boilers powered by methane are a major use for this material as a way to generate power.

10.2 Methanol

10.2.1 Production

Methanol may appear closely related to methane in terms of molecular weight and size, but the production of methanol is not often directly from methane in a single step. Methanol can actually be made from a wide variety of different feed stocks, but for our purposes here, we will concentrate on its production from what is called syn gas. The general steps in the process are as follows:
1. The feedstock needs to be made into syn gas – synthesis gas – which means CO and H_2 gases.
2. Methane and natural gas is often used for syn gas, but most plant materials can be used as a syn gas feedstock.
3. Two large steps are always involved: first, get feedstock to a syn gas stream of CO_2, CO, and H_2.
4. Next, catalysis converts the gas stream to methanol.
5. The reaction chemistry can be summarized as:
   - $3H_2O + 2CH_4 \longrightarrow CO + CO_2 + 7H_2$, the production of syn gas, and
   - $7H_2 + CO + CO_2 \longrightarrow 2CH_3OH + H_2O + 2H_2$.
6. The excess hydrogen gas can be re-cycled into the process to produce more methanol, if there is sufficient CO and CO_2 available.
7. Also, the production is exothermic enough that the heat generated can be used to generate power, electricity, which is needed in the process.

https://doi.org/10.1515/9783110671094-010

10.2.2 Uses

Syn gas is used to produce five major organic products, after the formation of methanol, all of which are then used in other, further applications. A simplified schematic is shown in Figure 10.1.

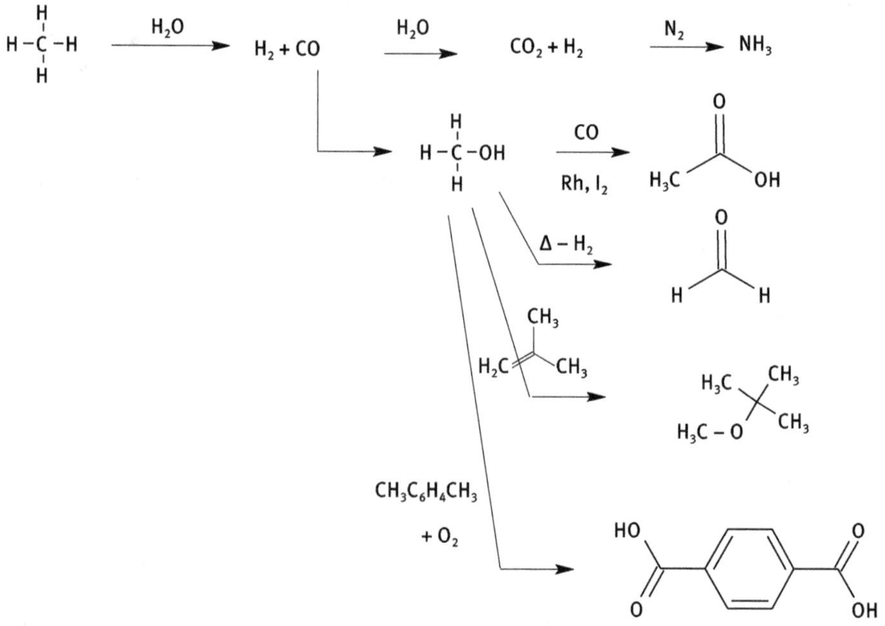

Figure 10.1: Syn Gas Production and Uses.

Each of these five chemicals is used in further reactions, as shown in Table 10.1. In the case of ammonia, it can be used for the production of further basic chemicals, or can be used directly (as fertilizer). Perhaps obviously, other components are required in each synthesis.

Table 10.1: Uses for Chemicals Made from Methanol.

Chemical from Methanol	Uses
Acetic acid	Vinyl acetate
Ammonia	Fertilizer
Formaldehyde	Resins
Methyl-tert-butyl ether	Fuel additive
Dimethylterephthalate	Production of PETE

In addition, methanol is used as an industrial solvent in several processes where a polar organic material is required [3].

10.3 Other oxygenated C1 chemicals

Beyond methane and methanol, the other two oxygenated C1 organics are formic acid and formaldehyde.

10.3.1 Production

Formaldehyde can be made on an industrial scale as shown in Figure 10.1, or from methanol and oxygen. In this case, a catalyst is required, which is usually a metal such as an iron oxide-molybdenum mix, or silver. The reaction is sometimes referred to as the Formox Process. The reaction chemistry is fairly straight forward, as follows:

$$O_2(g) + 2CH_3OH \longrightarrow 2H_2O + 2CH_2O \quad \text{at } 250\,°C\text{--}400\,°C$$

Silver catalysts require higher temperatures, approximately 650 °C, but are still economically viable [2].

Formic acid is made by the combination of methanol and carbon monoxide, with a strong base present, at roughly 40 atm and 80 °C. The immediate product is methyl formate, which is then treated with a base, such as sodium methoxide, with a large excess of water. The result is formic acid and methanol. The reaction chemistry can be represented as follows:

$$CH_3OH + CO \longrightarrow HCO_2CH_3$$

Followed by:

$$HCO_2CH_3 + H_2O \longrightarrow HCO_2H + CH_3OH$$

10.3.2 Uses

Formaldehyde is used in numerous resins and plastics, and is widespread enough that nearly 10 million tons are produced annually. The traditional use of formaldehyde for embalming has decreased in recent years because of concerns about long term effects of it in the environment. European Union countries stopped using it in this capacity in 2007.

Formic acid is produced on a scale of hundreds of thousands of tons per year, and is used in leather tanning, in animal feed as a preservative. A major end use for it has been as a toilet bowl cleaner.

10.4 CFCs and HFCs

The production of a class of gases and materials for refrigeration purposes is very closely linked with the C1 fraction, namely the chlorofluorcarbons and the hydrofluorocarbons, many of which contain only one carbon atom. We will treat the entire class of CFCs and HCFCs here, for the sake of completeness, even though they are not all directly produced from methane.

Even though the use of CFCs was banned in 1978, they still represent a class of heat exchange materials that are far more effective than the previous materials that had been used for refrigerants, such as ammonia and sulfur dioxide. In addition, DuPont research and development has worked to find more environmentally friendly materials that can perform as well as CFCs without causing environmental damage [1].

10.4.1 Production

The starting materials for many CFCs and HFCs are either chloroform or carbon tetrachloride. In turn, carbon tetrachloride is produced through the combination of methane with sulfur, then chlorine. The simplified reaction chemistry is shown in Scheme 10.1.

$$CH_4 + 4S \longrightarrow CS_2 + 2H_2S$$
$$CS_2 + 3Cl_2 \longrightarrow S_2Cl_2 + CCl_4$$

From CCl_4:

$$2CCl_4 + 3HF \longrightarrow CCl_2F_2 + CCl_2F + 3HCl$$

Scheme 10.1: CFC Production.

The hydrofluoric acid (HF) needed in the production of CFCs, above, is extracted from calcium fluoride with H_2SO_4 (another industrial use for sulfuric acid). As well, this represents another use of elemental chlorine.

10.4.2 Nomenclature

CFCs and HFCs are a class of compounds that use a nomenclature that is neither IUPAC nor American Chemical Society approved. The nomenclature was developed by the Dupont Company, and is still used by many today. It follows a few simple rules. They are:
1. Each CFC or HFC name has three numbers.
2. If the first number is 0, then it is dropped from the name.
3. First digit = carbon atoms −1.
4. Second digit = hydrogen atoms +1.
5. Third digit = fluorine atoms.

6. When two carbon isomers are possible, the symmetrical isomer has no further iden-
    tifier. As isomers become less symmetrical, letters, starting with 'a,' are used.

Examples of this include:

- CCl_2F_2 — CFC-12 (the first digit is 0 and is dropped)
- CCl_2FCCl_2F — CFC-112
- CCl_3CF_3 — CFC-113b (the 'b' indicates the least symmetrical placement of the
  chlorine and fluorine atoms on the molecule)

10.5 Elemental hydrogen production from methane

Hydrogen stripping from the lightest fraction of crude oil or from natural gas yields
enough hydrogen gas that it is combined with elemental nitrogen to produce much of
the world's synthetic fertilizer, as discussed in Chapters 4 and 17.

Steam reforming to produce hydrogen can be shown with a simple chemical reaction:

$$CH_4(g) + H_2O(g) \longrightarrow 3H_2(g) + CO(g)$$

Reaction conditions are generally 700–1,100 °C, and 20 atm. This reaction is simply the
syn gas reaction shown in Figure 10.1, but stopped at the point where hydrogen can be
recovered. Methane does not have to be used as the starting material for this reaction (as
seen in Chapter 7 when discussing the production of carbon black), but it is very often
economically feasible to use methane.

The other means by which hydrogen gas is produced is the chlor alkali process,
which was discussed in Chapter 6. In this case, hydrogen is considered the third product,
after chlorine and sodium hydroxide.

10.6 Pollution and recycling

As was seen, methanol is almost entirely consumed in the production of further basic
chemicals. Thus, none is recycled.

Additionally, formic acid and formaldehyde are used in further reactions, or as end
use products, and are thus consumed. Neither of these is recycled.

Since chlorofluorocarbons and hydrofluorocarbons were routinely are used in consumer products, recycling them was a matter of disassembling the end product and recycling components. Unfortunately, many times a product is discarded because the CFC
or HFC had leaked out of it. Thus, these materials dispersed to the atmosphere.

Bibliography

[1] Dupont. Website. (Accessed 29 November 2022, at https://www.dupont.com).

[2] The Formox Process, Perstorp. Website. (Accessed 29 November 2022, at www.perstorp.com/en).

[3] The Methanol Institute. Website. (Accessed 29 November 2022, at www.methanol.org/).

11 The C2 fraction and ethylene chemistry

11.1 Introduction

The reactivity of the double bond dominates the chemistry of the C2 fraction. While there are several polymers created from the opening of an olefinic bond, such as polypropylene, polyvinylchloride, polyacrylonitrile, and polystyrene – all of which will be treated here or in other chapters – the simplest double bond chemistry is that of ethylene. The production of polyethylene from ethylene, shown in Figure 11.1 in a simplified manner, has been examined in enough detail that reaction conditions have been established which routinely produce what is termed "high density polyethylene," or HDPE, and "low density polyethylene," or LDPE [1]. In the last thirty years, there have been significant advances in this field such that there are now manufacturing concerns which produce even ultra-high molecular weight polyethylene and very-low density polyethylene [1]. But ethylene is also used to produce a series of basic chemicals, which we will discuss here.

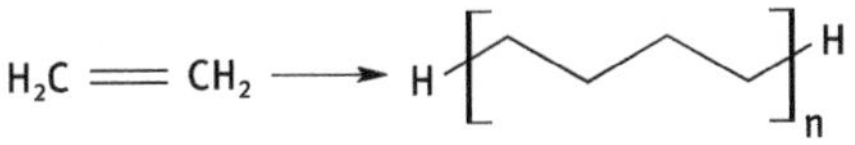

$$H_2C = CH_2 \longrightarrow$$

Figure 11.1: Polyethylene Manufacture.

The basic commodity chemicals produced from ethylene are shown in Figure 11.2. From these seven, there are several further basic chemicals produced, which are listed in Table 11.1. One can also claim that ethylbenzene, and thus styrene, are made from ethylene. We will treat these two chemicals under the discussion of organics, because they also require benzene for their production.

11.2 Ethane

One of the minor components of natural gas, ethane can be used as a fuel source, especially when still mixed with methane, but is often separated cryogenically. It can then be dehydrogenated to ethylene, to meet the large need for this material as a starting monomer for polyethylene of all densities.

11.3 Ethylene

Ethylene is one of the major basic chemicals that are the targets of petroleum cracking, the process of breaking larger molecular weight hydrocarbons in crude oil down to smaller, more economically useful ones.

https://doi.org/10.1515/9783110671094-011

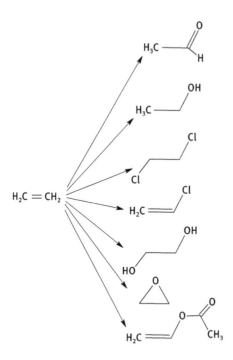

Figure 11.2: Basic Chemicals Produced from Ethylene.

Table 11.1: Basic Chemicals Produced from the Immediate Chemicals Made from Ethylene.

Ethylene Derivative	Further Basic Chemicals
Acetaldehyde	Acetic acid
	Acetic anhydride
Ethanol [2]	Ethyl ether
	Chloral
	Ethyl bromide
	Ethyl amine
Ethylene dichloride	Vinyl chloride
	Polychloroethanes
Vinyl chloride	Polyvinyl chloride (PVC)
Ethylene glycol	Polyesters
Ethylene oxide [4]	Polyethyelene oxide (PEO)
	Ethylene glycol
Vinyl acetate	Polyvinyl acetate
	Polyvinyl alcohol

It may be fair to say that the use of ethylene to make polyethylene, in all its many forms, is one of the greatest chemical changes mankind has ever affected in the world. The mass production of polyethylene and formation of it into useful products and materials may rank with the domestication of animals, the advent of farming, and the invention of the wheel as one of the most important improvements that human history has seen. But polyethylene is not the only polymer that requires the ethylene unit to proceed from monomer to polymer. Polypropylene, polyvinyl chloride, polybutadiene, and polystyrene all require the olefin double bond for their respective polymerizations. Polyethylene is however, the polymer with the simplest repeating chain unit.

$H_2C = CH_2 \longrightarrow$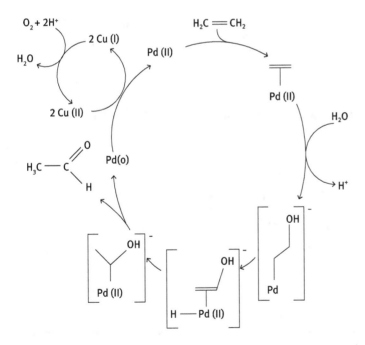

11.3.1 Acetaldehyde and acetic acid

These two chemicals are always connected in an industrial setting, because acetaldehyde is used to make acetic acid via the Wacker Process, as seen in Figure 11.3. Ethylene is added to a palladium (II) catalyst, usually $PdCl_2$, with water, and the palladium is reduced to Pd(0), palladium metal, as the acetaldehyde is produced. To be catalytic, the palladium must be regenerated as Pd(II), and this is done with a copper salt as a concurrent oxidation–reduction.

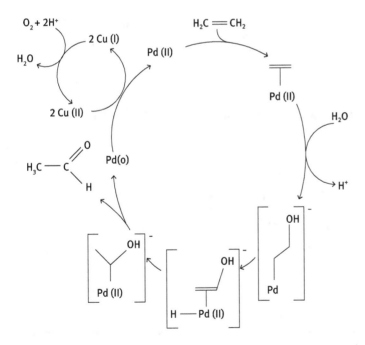

Figure 11.3: The Wacker Process.

Almost all acetaldehyde is used to manufacture acetic acid, although today acetic acid can also be made from methanol. The carbonylation of methanol with an iridium catalyst is called the Cativa Process, and its carbonylation with a rhodium catalyst is called the Monsanto Process.

A large portion of acetic acid is used to produce vinyl acetate, while the rest of it is split between uses in manufacture of esters and acetic anhydride, as well as use as a polar organic solvent.

Esterification of acetic acid with different alcohols will produce an array of esters, although the production of ethyl acetate from acetic acid and ethanol is widely used in various paints and inks.

The production of acetic anhydride is a large scale operation that simply requires the dehydration of acetic acid, as shown:

$$2CH_3CO_2H \longrightarrow (CH_3CO)_2O + H_2O$$

This reaction generally proceeds at 700 °C. The major use for acetic anhydride is in the production of cellulose acetate, a synthetic fiber.

11.3.2 Ethylene dichloride (also, ethylene chloride, or 1,2-dichloroethane)

Millions of tons of ethylene dichloride are produced annually from the reaction of ethylene and chlorine (use for chlorine) using $FeCl_3$ as a catalyst. The reaction chemistry can be represented as:

$$CH_2CH_2 + Cl_2 \longrightarrow ClCH_2CH_2Cl$$

which occurs at 50 °C, and looks like a direct addition reaction, although it will not occur without iron (III) chloride.

A second type of production method, the oxychlorination of ethylene, requires hydrochloric acid and $CuCl_2$, according to the reaction:

$$2CH_2CH_2 + O_2 + 4HCl \longrightarrow 2ClCH_2CH_2Cl + 2H_2O$$

Most ethylene dichloride is consumed in the production of vinyl chloride, with some firms co-producing each material.

11.3.3 Vinyl chloride

Thermal dechlorination of ethylene dichloride is a direct route to vinyl chloride. The reaction chemistry is:

$$ClCH_2CH_2Cl \longrightarrow CH_2=CHCl + HCl$$

The by-product HCl must be removed before further use of the product.

Vinyl chloride can also be produced at 20–30 atm and 500 °C, with hydrochloric acid as a by-product.

$$ClCH_2CH_2Cl \longrightarrow CH_2CHCl + HCl$$

Because of the temperature at which this reaction must be carried out, there are also other chlorinated hydrocarbon by-products. These, as well as the HCl generated, must be separated from the product before it can be further used.

All vinyl chloride is used in the production of polyvinyl chloride [3].

11.3.4 Ethylene oxide and ethylene glycol

Ethylene oxide was once produced by the chlorohydrin process, although this has now been replaced by the direct oxidation of ethylene with a silver catalyst. The chlorohydrin process is still used in the production of propylene oxide, and will be discussed in the next chapter.

The direct oxidation of ethylene to ethylene oxide was originally patented in 1931, although its use as the major process in industry took a further several years. It can be represented as:

$$CH_2 = CH_2 + O_2 \longrightarrow Ag \longrightarrow CH_2CH_2O$$

When it comes to use, a large percentage of ethylene oxide is consumed to make ethylene glycol through an acid catalyzed ring opening, or heat catalyzed ring opening, with water. The reaction chemistry for this is represented as:

$$CH_2CH_2O + H_2O \longrightarrow HOCH_2CH_2OH$$

Beyond this, ethylene glycol is used in anti-freeze (freezing point –45 °C), which everyone seems to know, but also used in synthetic, polyester fibers, and in ethanolamines. It's use in the production of the plastic polyethylene terephthalate will be discussed in Chapter 15.

11.3.5 Vinyl acetate

Vinyl acetate is made on a large scale from acetic acid, and the reaction requires a palladium catalyst, much like Wacker chemistry. The reaction can be shown as:

$$CH_2=CH_2 + CH_3CO_2H + \frac{1}{2}O_2 \longrightarrow CH_3CO_2CH=CH_2 + H_2O$$

Some over oxidation can occur in this reaction, producing CO_2.

Vinyl acetate can be polymerized to polyvinyl acetate (sometimes abbreviated PVA, although this can also be used to designate polyvinyl alcohol). As well, it can be co-polymerized with other vinyl monomers for a variety of uses.

Vinyl acetate also finds consumer end uses in adhesives, paints, and other types of coatings.

11.3.6 Ethanol

Of the seven basic chemicals we have listed as directly produced from ethylene, ethanol is the only one that is also produced in another fashion. The truly classic, time-honored method of ethanol production is fermentation. This involves yeast causing the break-down of carbohydrates in plant matter, and virtually every cereal grain or sugar ever cultivated has been used to make alcohol and alcoholic beverages. The topic has been the subject recently of more than one excellent book [5]. The reaction chemistry is as follows:

$$C_6H_{12}O_6 \longrightarrow 2C_2H_5OH + 2CO_2$$

Natural fermentation does not produce ethanol to greater than about 15 %, the point at which most yeast dies in the alcohol it has produced. Distillation is required to enhance the alcohol content (for example, the distillation of wine produces brandy). Industrial distillation can concentrate alcohol to almost 96 %. Purification beyond this requires an additive to entrain the remaining water in the mixture.

Industrially today, ethanol is produced as follows:

$$CH_2CH_2 + H_2O \longrightarrow CH_3CH_2OH$$

The process is acid catalyzed, generally using phosphoric acid [4], at approximately 300 °C. Silica gel is required as a porous support for the reaction.

Uses as a fuel

Bioethanol has recently been used as a fuel in several countries. What is called E5 (mean-ing 5 % ethanol) is used in cars in the USA now, and there is discussion about moving to E10. Beyond this, E85 gas stations exist. It has been found that E85 movement through pipes is more corrosive than moving traditional hydrocarbon gasoline.

In the United States, the predominant source of material for bioethanol is corn. In Brazil, another large user of bioethanol, it is sugar cane. There has been considerable work in the past ten years about producing bioethanol from cellulosic material, but none has yet been commercialized [2].

Other uses

Beyond fuel, much of the ethanol produced is used as a solvent. Because it is readily miscible with water, it can be found in numerous personal care products. Additionally, a small amount of ethanol is used medically to treat poisoning by methanol and ethylene glycol.

11.3.7 Acetylene

Acetylene is currently made from methane combustion. It is essentially the only two-carbon atom molecule not made from ethylene. It is also a secondary product in the production of ethylene, during the cracking of heavier hydrocarbons.

Uses of acetylene as a welding feed gas are well known, but it was once also used in the production of vinyl acetate and the production of acrylic acid. Both of these techniques are no longer large production processes.

Several hundred thousand tons of acetylene are still made annually. This is used for welding operations.

11.4 Recycling

The recycling associated with any C2 material is almost always that of the recycling of a finished plastic and not of a monomer unit or of a basic chemical. Exceptions to this might be ethylene glycol used as anti-freeze, and ethanol.

There is interest in the recycling and re-use of ethanol as an industrial solvent, but there have not yet been large scale recycling operations developed, or companies specifically dedicated to the recycling of it.

The other basic chemicals discussed here are all consumed in the production of further chemicals or end use products, as shown in Table 11.1.

Bibliography

[1] American Chemistry Council. Website. (Accessed 30 November 2022, at https://www.americanchemistry.com).
[2] Ethanol Producer Magazine. Website. (Accessed 30 November 2022, at www.ethanolproducer.com/).
[3] European Council of Vinyl Manufacturers (ECVM). Website. (Accessed 30 November 2022, at www.pvc.org/).
[4] Shell Chemical. Website. (Accessed 30 November 2022, at https://www.shell.com/service/search.html#q=ethylene%20oxide).
[5] Standage, Tom. A History of the World in Six Glasses, Walker and Company, New York, 2005.

12 C3 and C4 fraction chemistry

12.1 Introduction

We have just seen that the chemistry of the carbon – carbon double bond dominates the reactivity of the C2 fraction, and the physical properties of the resulting materials. The means by which the bond activation was initiated, including choice of catalysts, heat, pressure, and specific pH, often dictates whether a resulting polyethylene is considered low density or high density, and what the physical characteristics are for any ethylene or polyethylene derivative. This double bond reactivity also dictates much of the chemistry of the C3 and C4 fraction, but there are other factors that become very important when considering the final physical properties of the resulting polymers that can be made by polymerizing the double bond when a methyl or other substituent is attached, pendant to the main chain.

The direction of any side chain generally falls into one of three possibilities when there are numerous repeat units: isotactic, in which all side chains are facing in the same direction from the main chain, syndiotactic, where the side chains alternate sides from one polymerized monomer unit to the next, and atactic, where there is no regularity to the side chain placement. We will discuss this in some detail throughout this chapter, as well as the basic chemicals that can be produced from the C3 and C4 stream. A more full discussion of polypropylene, which accounts for roughly 65 % of propylene use, is in Chapter 15.

12.2 Propane

A significant amount of propane, the only saturated C3 carbon compound, is simply burned for heating fuel in developed and developing countries throughout the world. It is a common enough fuel that trade organizations, such as the National Propane Gas Association, or the Canadian Propane Association, exist to promote its use, and its continued safe handling [3, 2]. Additionally, the Propane Education and Research Council promotes the safe use of this material in a variety of situations and applications, as does the organization titled Propane 101 [6, 5]. The World Liquid Petroleum Gas Association also tries to educate the public about the uses of propane, but perhaps obviously, concerns itself with all materials that are part of natural gas [7].

A growing use for propane has been as the main fuel in light motor vehicle transportation, since it is a clean burning fuel, without a large percentage of partially oxidized carbon products after combustion, and because it can be liquefied at relatively low pressure, making it a convenient material for the "gas tank." When used in homes for heating, or for cooking food, a small amount of what is called an "odorant" – a molecule that can be smelled in very small concentration – is often added. Ethanethiol (C_2H_5SH) is often used in this regard because of its distinct, unpleasant aroma. The reason for adding

https://doi.org/10.1515/9783110671094-012

an odorant is a personal safety consideration, so that any leaks in the propane gas line in the home can be detected before a large enough build-up occurs that the home or commercial building could explode and burn.

12.3 Propylene

Propylene exists as a smaller percentage than ethylene in the separable products from the light fraction of crude oil, and can be extracted from coal if necessary. Some of it is always used to alkylate gasoline mixtures. This portion is normally not reported as a separate, isolated product by the companies that utilize it. However, a large portion of propylene is made by steam cracking of hydrocarbons, just as ethylene is. The reaction chemistry can be represented rather simply, as shown in Scheme 12.1, as follows:

$$2C_3H_8 \longrightarrow H_3C-CH=CH_2 + H_2C=CH_2 + CH_4 + H_2$$

Scheme 12.1: Propylene production.

On an industrial level, the cracking does not give a single, pure product as Figure 12.1 implies. But product separation has become a developed part of the petroleum refining industry over the course of decades, and all the products are ultimately used. Propane is not the only feedstock for the production of propylene. Other fractions, such as gas oil or naphtha, can be used as well. The choice of feedstock is routinely one of price.

Propylene is used to make other small organic products, or is used to make various types of polypropylene. Figure 12.1 shows the major small, organic chemicals produced from propylene.

As can be seen from Figure 12.1, propylene is used to produce propylene oxide, propylene chlorohydrin, acrylonitrile, isopropanol, acetone, and phenol. Of course, other carbon-containing molecules must be incorporated in these processes.

12.4 Acrylonitrile

The current method of acrylonitrile production has been utilized for the past forty years, and also produces acetonitrile as a by-product, which is another marketable chemical. Called the Sohio Process, this becomes another use for ammonia, which was discussed in Chapter 4. Scheme 12.2 shows the reaction chemistry:

The reaction proceeds at elevated temperatures and pressures (400–500 °C, 50 atm with a $Bi_2O_3 \cdot MoO_3$ catalyst). While the reaction appears straightforward, this also produces acetonitrile, and emits hydrogen cyanide, CO and CO_2, all of which are either captured for re-use, or captured to prevent environmental degradation.

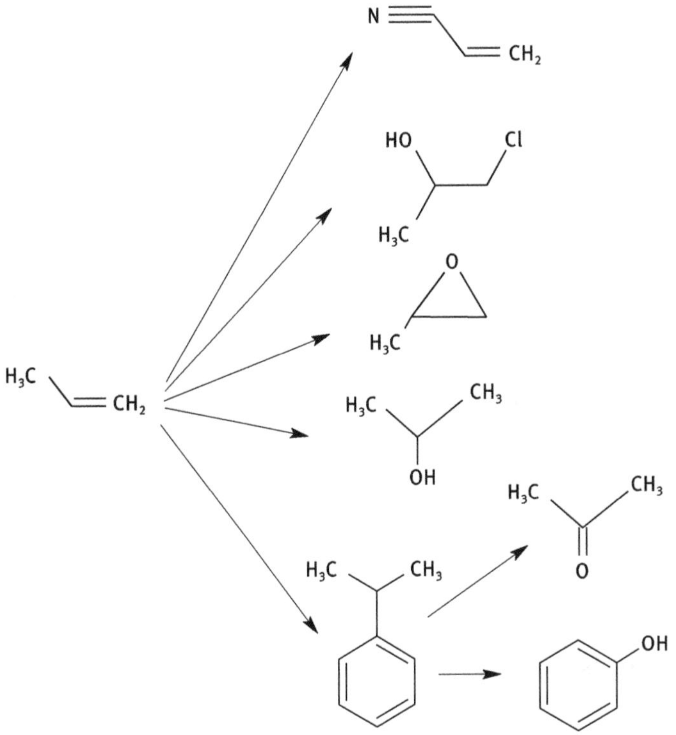

Figure 12.1: Uses of propylene.

$$2NH_3 + 3O_2 + 2CH_2CHCH_3 \longrightarrow 2CH_2=CH - CN + 6H_2O$$

Scheme 12.2: Acrylonitrile production from propylene.

There are several smaller uses of acrylonitrile, but the polymerization of it to form polyacrylonitrile as a homo-polymer is by far the predominant one. Polyacrylonitrile is routinely used in the production of synthetic fabrics, which are used in a variety of applications and consumer products [4].

12.5 Propylene oxide and propylene chlorohydrin

It was mentioned in Chapter 11 that the chlorohydrin method for the manufacture of ethylene oxide has been phased out by more efficient methods. For the production of propylene oxide though, it is still one of the reactions of choice for producing propylene oxide. Methods of propylene oxide production include the following, in Scheme 12.3.

The mixture of the two chlorine-containing isomers does not pose a problem, as a base is used to produce the propylene oxide, as follows in Scheme 12.4.

$$2CH_2CHCH_3 + Cl_2 + H_2O \longrightarrow CH_3CHClCH_2OH + CH_3CHOHCH_2Cl$$

Scheme 12.3: Production of Chlorohydrin.

$$CH_3CHOHCH_2Cl + OH- \longrightarrow CH_3CHCH_2O + Cl - +H_2O$$

Scheme 12.4: Production of Propylene Oxide.

The chlorine is removed as NaCl or CaCl$_2$ and the propylene oxide isolated as a final step.

In 2009, a new approach came on line in which hydrogen peroxide is reacted with propylene directly, to produce only propylene oxide and water. The reaction chemistry can be represented as is shown in Scheme 12.5.

$$CH_3CHCH_2 + H_2O_2 \longrightarrow CH_3CHCH_2O + H_2O$$

Scheme 12.5: Exclusive Propylene Oxide Production.

The major benefit of the process is the lack of any co-products or by-products.

The majority of propylene oxide is used to produce polyurethanes. A significant portion is used for the production of propylene glycol.

12.6 Isopropanol

There are several processes by which isopropanol (sometimes still called rubbing alcohol) can be produced. They include the hydration of propylene with water, and the hydrogenation of acetone. In general, the reaction is shown in Figure 12.2.

Figure 12.2: Isopropanol Production.

Indirect hydration involves the use of sulfuric acid (another use of this product, which was discussed in Chapter 2), and results in the alcohol and regeneration of the acid upon addition of water. The advantage of this approach is that relatively impure propylene can be used as a feed. The reaction chemistry is shown in Figure 12.3.

Direct hydration requires an elevated pressure and a higher purity of propylene feed (generally, greater than 90 %), but does not require the use of sulfuric acid.

In both forms of hydration, the resulting alcohol must be distilled from water to isolate it.

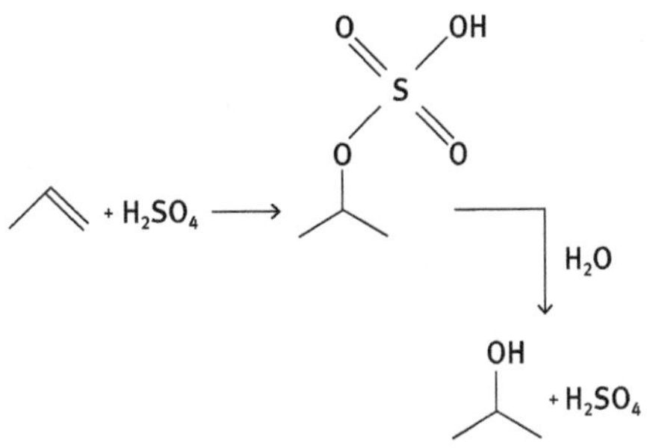

Figure 12.3: Isopropanol Production with H_2SO_4.

Acetone hydrogenation requires a metal catalyst, usually a copper – chromium oxide catalyst, or sometimes nickel.

Isopropanol is widely used as a solvent, but is also used as a cleaning fluid. Per year, tens of thousands of tons are produced for use as a solvent because of its ability to dissolve a wide array of organic materials, and because it is relatively non-toxic when compared with other solvents of like polarity.

12.7 Cumene

Cumene is produced through the reaction of benzene and propylene in a Friedel-Crafts alkylation. Alumina was the catalyst of choice for this reaction for years, but has been displaced in the past ten years by zeolite catalysts. The reaction chemistry appears straightforward, and can be represented as follows, in Figure 12.4.

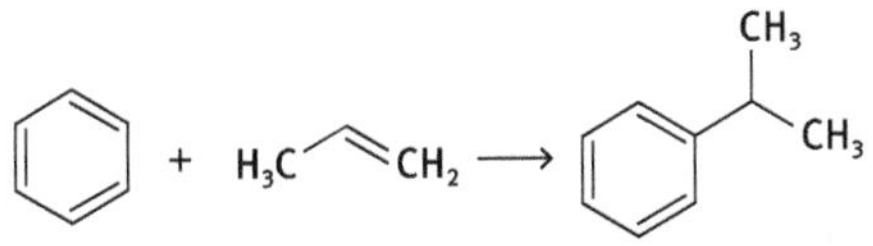

Figure 12.4: Cumene Production.

Almost all cumene is used in the production of acetone and phenol. Only small amounts are used as organic solvents.

12.8 Acetone and phenol

These two chemicals may not seem connected, but they are always co-produced from cumene in what is called the Cumene Process, sometimes called the Cumene-Phenol Process, or the Hock Process. This has been the synthetic pathway for both chemicals since the 1940s, which produces phenol on the scale of several million tons per year.

This reaction is run at elevated temperature and pressure (250 °C and 30 atm), and does require an acid catalyst. Phosphoric acid is often used because it is very inexpensive. The catalyst makes a stable radical, which can then bond to oxygen gas. The subsequent rearrangement, called the Hock rearrangement, forms the aromatic ring to oxygen atom bond. The addition of water then provides the oxygen needed for the material to separate into phenol and acetone.

The two chemicals are then distilled to effect their separation. The reaction chemistry is shown in Figure 12.5.

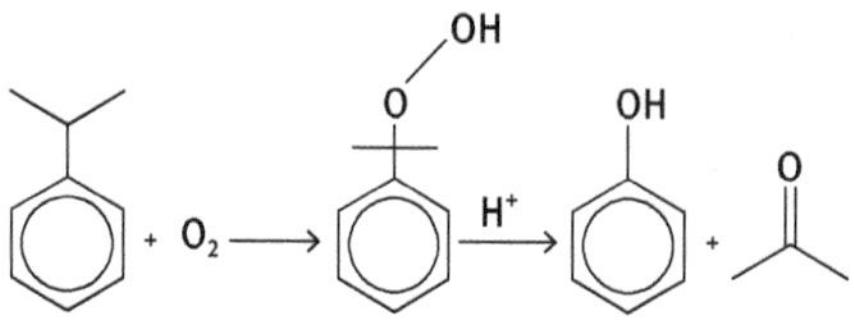

Figure 12.5: Production of Acetone and Phenol.

12.9 Butane

As with propane, the saturated hydrocarbon butane is used directly by consumers as a fuel. Butane cigarette lighters are ubiquitous in most countries, although larger quantities of butane are often sold for a variety of other uses.

12.10 The 1,3-butadiene isomers

Synthetic rubber requires the use of 1,3-butadiene, and thus this is obtained in large quantities from the C4 fraction of petroleum. Steam cracking, which we have mentioned as a method to produce ethylene and other olefinic small molecules, also produces 1,3-butadiene, especially when heavier hydrocarbons are used as the feedstock. As mentioned, the steam cracking process requires the addition of steam at elevated temperatures (roughly 900 °C) to effect the loss of hydrogen atoms from saturated small hydrocarbons.

Butadiene can also be obtained by the dehydrogenation of butane, a process that is effected catalytically. This process is not the major method of production of the diene, but it has been in use since the Second World War, as a first step in the production of synthetic rubber, as shown in Figure 12.6.

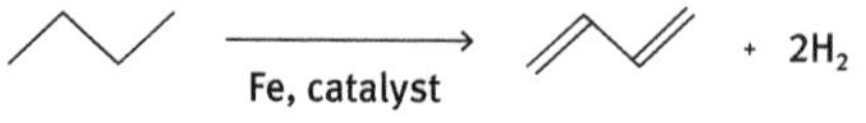

Figure 12.6: Butadiene Production from Butane.

In addition, metal oxide catalysts can be used to produce butadiene from ethanol. The method has been used throughout the world in the past, but is not currently used on a large scale in the United States, Canada, or Europe.

Uses

Almost all butadiene is used in the manufacture of synthetic rubber. Natural rubber, which is discussed in Chapter 21, has isoprene as its monomeric unit, which is only one methyl group different from butadiene. Synthetic rubber is often made with some form of co-polymerization of butadiene and another polymer or polymer blend. Several million tons are produced annually, although this is the sum of several different formulations [1]. A few examples from the many types are shown in Table 12.1.

Table 12.1: Butadiene Rubber Types.

Co-polymer with Butadiene	Example uses
Polychloroprene	Neoprene textile
Acrylonitrile	Synthetic textiles, clothing
Acrylonitrile/butadiene/styrene (ABS)	Automotive body components
Acrylonitrile/butadiene (NBR)	Medical uses
Styrene/Butadiene (SBR)	Automobile tires

It can be seen that the use of SBR is favored for automobile tires, but there are other formulations as well as natural rubber that can be used.

12.11 Recycling possibilities

The recycling of the basic starting materials discussed in this chapter never occurs, simply because all of these materials are consumed in the production of other materials, the latter often being some consumer product. Much like the materials discussed in Chapter 11, the basic chemicals produced from the C3 and C4 fraction are entirely consumed in further processes. The recycling of rubber is not yet a large scale operation, but does occur in some areas. It is discussed in Chapter 21.

Bibliography

[1] American Chemistry Council. Website. (Accessed 2 December 2022, at https://www.americanchemistry.com/chemistry-in-america/chemistry-in-everyday-products/plastics).
[2] The Canadian Propane Association. Website. (Accessed 2 December 2022, at www.propane.ca/).
[3] National Propane Gas Association. Website. (Accessed 24 December 2022, at https://www.npga.org/).
[4] Plastics Industry Trade Association. Website. (Accessed 2 December 2022, at www.plasticsindustry.org/).
[5] Propane 101. Website. (Accessed 2 December 2022, at www.propane101.com/).
[6] Propane Education and Research Council. Website. (Accessed 2 December 2022, at https://propane.com/).
[7] World LP Gas Association. Website. (Accessed 2 December 2022, at https://www.wlpga.org/).

13 Liquid organic fuels

13.1 Gasoline

The C1 to C4 fraction of crude oil has been covered in Chapters 10–12. Beginning with *n*-pentane and moving to progressively higher molecular weight hydrocarbons, the saturated hydrocarbon molecules exist as liquids, then eventually as waxy solids. The most common liquid fuel to the average person is undoubtedly gasoline, which is a mixture of hydrocarbon isomers that generally contain seven or eight carbon atoms, and exists as a non-viscous, flammable liquid [1].

13.2 Octane number

Determining what is called the octane number is a measure of reliably figuring how well a particular fuel blend burns. Despite the name "octane," what makes up gasoline is not a cleanly distilled fraction of the molecule *n*-octane. Indeed, *n*-octane burns rather poorly, while branched hydrocarbon isomers combust much better.

When the octane rating system was developed, the isomer 2,2,4-trimethypentane was assigned the number 100. The molecule *n*-heptane was assigned the number 0. The amount of branching in hydrocarbons generally improves the performance of the fuel. Figure 13.1 illustrates these two molecules. Both of these start points were used to determine a scale for anti-knock properties for gasoline blends. Blends of the two compounds were used to determine the points in between 100 and zero. Beyond this, any gasoline blend can be compared to the scale, and a rating assigned.
- 100 – 2,2,4-trimethylpentane
- 0 – *n*-heptane

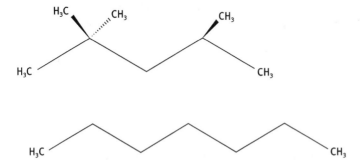

Figure 13.1: 2,2,4-Trimethylpentane and *n*-Heptane.

Determining the current number for motor fuel found at gas stations is a matter of combining two numbers, the Research Octane Number (RON), and the Motor Octane

https://doi.org/10.1515/9783110671094-013

Number (MON). The RON is based on a motor functioning at 600 rpm and 125 °F, which is a generally low speed. The MON is computed based on a motor functioning at 900 rpm and 300 °F, a generally high speed. Now numbers at the pump are reported as an average of the two numbers, since automobile use can be widely different, depending upon the driving circumstances. Thus, (R + M)/2 is reported.

13.3 Additives

Tetraethyl lead was used decades ago because it did improve engine efficiency and reduce knocking. Only 1–5 mL of the additive was used per gallon of gasoline. But it has been completely phased out because of the human and environmental damage it causes, and has been replaced by a variety of different materials. These include alcohols, ethers, and several antiknock agents and stabilizers. In addition, fuel dyes can be added for fuels designed specifically for small engines.

In the United States, 5 % ethanol is routinely added to motor gasoline, and a debate is underway concerning bringing this up to a 10 % ethanol blend. European Union countries also sell and use gasoline with ethanol added, although not to exactly the same percentage as E5.

The sale of automobiles that burn ethanol have become large enough that filling stations do exist that sell E85, meaning an 85 % ethanol fuel.

The list of additives has become quite long as the use of automobiles and light trucks has become more ubiquitous, and their use has spread to more and more extreme environments. A current, but not all inclusive list is shown in Table 13.1.

Beyond additives, many fuels are blended for use in different environments. The average automobile with an internal combustion engine does not travel more than 300 miles (450 km) without refueling, so the blend of fuel does not change appreciably in that distance. Some vehicles however, must be used far from where they are normally based, such as heavy construction equipment and military vehicles. In these cases, blends of gasoline and diesel may be required so that fuel still burns in extreme low or high temperatures. In very low temperature winters in remote areas, engines are sometimes simply turned on and kept running for the winter (an example might be the Russian far east).

13.4 Jet fuels

Gas-turbine engines used in civilian and military aircraft require certain formulations of liquid fuels which are generally called jet fuel. Like gasoline, these fuels are a mixture of isomers of hydrocarbons, usually in a molecular weight range from 9 to 16 carbon atoms, generally called kerosene when used in other applications.

Table 13.1: Gasoline Additives.

Class of Additive	Material	Reason
Alcohols	Ethanol	Antiknock
	Methanol	Antiknock
	Isopropyl alcohol	Antiknock
	n-Butanol	Antiknock
Ethers	t-Amyl methyl ether	Starter promoter
	t-Amyl ethyl ether	Starter promoter
	Ethyl-t-butyl ether	Starter promoter
	t-Hexyl methyl ether	Starter promoter
	Di-isoprpyl ether	Starter promoter
Hydrocarbons	Iso-octane	Antiknock
	Toluene	Antiknock
Dyes	Blue 25	Fuel identifier
	Red 24	Fuel identifier
	Red 26	Diesel fuel dye
	Yellow 124	Diesel fuel for heating
Silicone		Anti-foaming agent
Amines	p-Phenylene diamine	Anti-oxidant
	Ethylene diamine	Anti-oxidant

Jet fuels are required to meet the same specifications from one country to another, to help make international travel by plane safer and more reliable. Thus, there are only a limited number of formulations and designations for such fuels, all of which have higher flash points than conventional motor fuel. The designators commonly used are: Jet A and Jet A-1, which are considered kerosene-type fuels with 8–16 carbons in their mixture. As well, Jet B is used in colder weather applications, and is sometimes called naphtha-type fuel. The carbon atom range for Jet B is 5–15 atoms. It is used almost exclusively by the military. The differences in freeze points and uses for these jet fuels are shown in Table 13.2.

Table 13.2: Jet Fuel Characteristics.

Name	Composition	Freeze Point	Use
Jet A		−40 °C	US Aircraft
Jet A-1		−47 °C	International, anti-static additive
Jet B	30 : 70 kerosene : gasoline	−60 °C	United States, some military

Since the Second World War, the United States military has developed a series of jet fuels for specific military applications, including aircraft at sea, and high speed aircraft. Designated JP and a number (such as JP-4), we mention them here, but note that they are not made on a large enough scale, or used in enough applications, that they are seen by most chemists and chemical engineers in normal industrial operations.

13.5 Diesel

Diesel fuel is used in a variety of individual and commercial vehicles, as well as military vehicles and equipment. The common factor in such vehicles is a diesel engine, an internal combustion engine that can burn the fuel without a spark plug.

While diesel produced from some renewable source is often titled biodiesel, the bulk of the diesel fuel consumed in the world remains that based on petroleum as a feedstock. This is now often called petro-diesel.

To produce petro-diesel, crude oil is fractionally distilled, with the 200–350 °C fraction producing a mixture of hydrocarbon molecules that normally contain 8–21 carbon atoms. Roughly $\frac{3}{4}$ of diesel is saturated hydrocarbon molecules in this molecular weight range, while the remainder is composed of aromatic hydrocarbons.

While gasoline is often characterized by a RON-MON average, diesel fuel is characterized by its cetane number. A higher cetane number indicates a fuel mixture that, when sprayed into heated, compressed air, ignites more easily.

In cold weather, diesel is often blended with small amounts of gasoline, because the raw diesel begins to wax and solidify. This increase in viscosity can clog fuel lines and render vehicles inoperable.

13.6 Liquefied petroleum gas

Liquefied petroleum gas (LPG) is a compressed mixture of propane and butane that can be used as fuel. While the volumes produced are not as great as for gasoline and diesel, LPG operations are still large enough that there are trade organizations for it. The Gas Processors Association states that it is organized around what is called mid-stream operations, and claims on its web site that: "Midstream operations include gathering, compression, treating, processing, marketing and storage of natural gas, as well as fractionation, transportation, storage and marketing of natural gas liquid" [2].

LPG has a long history of use in space heating and in cooking, but of late has become more of an alternative in terms of a motor fuel. As mentioned in Chapter 12, propane has become a motor vehicle fuel for applications where a cleaner exhaust stream is required.

13.7 Biofuels, pollution, and recycling

Since gasoline, jet fuel, and diesel fuel are produced from crude oil, a non-renewable material source, there have been numerous attempts to produce these fuels from some bio-based matter. In theory, using some plant or animal source for hydrocarbon based fuels will ultimately be carbon neutral, which means that any emitted carbon dioxide will be re-absorbed when the next generation of plants is grown [3].

Current problems with the production of biofuels are: the cost of biofuels when compared to traditional petroleum, land use for biofuel production, and the competing use of feedstock for biofuels and for food. In the United States, the feedstock for most bio-ethanol fuel is corn, which is also used in roughly $1/3^{rd}$ of the food products sold in a supermarket (as the ingredient corn syrup). Biodiesel is produced from a variety of plant sources, but can be produced from animal sources, including chicken fat.

In Brazil, the feedstock for biodiesel is soybeans, which is also used in numerous foods and food products [4]. Currently, just over 7 % of Brazil's soybeans are grown for the production of biodiesel [4]. Additionally, Brazil has seen the loss of a significant amount of the Amazon rain forest, especially over the past thirty years, as land is cleared to raise soybeans.

Additionally, from a purely chemical point of view, the production of biodiesel involves transesterification to produce fatty acid esters, which also produces glycerol, as shown in Figure 13.2. The excess capacity of glycerol on the market has severely driven its price down in recent years.

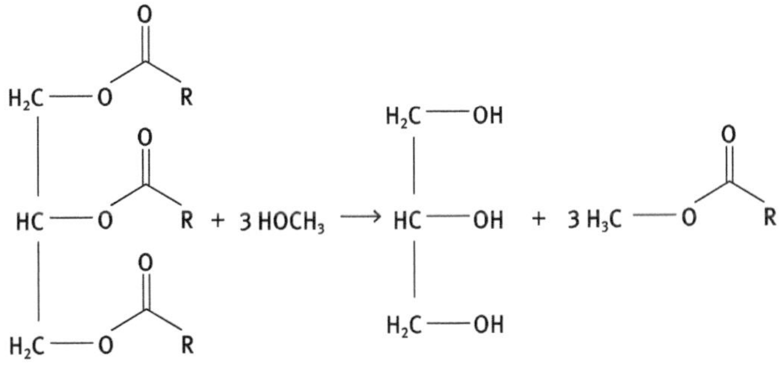

Figure 13.2: Transesterification of Fatty Acids.

Bibliography

[1] OPEC, Organization of the Petroleum Exporting Countries. Website. (Accessed 2 December 2022, at https://www.opec.org/opec_web/en/).
[2] Gas Processors Association. Website. (Accessed 2 December 2022, at: https://www.spglobal.com/engineering/en/products/gpa-standards.html).
[3] Fram Energy. Website. (Accessed 2 December 2022, as: https://farm-energy.extension.org/animal-fats-for-biodiesel-production/#:~:text=Waste%20fat%20from%20animal%20carcasses,separates%20and%20pathogens%20are%20destroyed).
[4] Soybean & Corn Advisor. Website. (Accessed 2 December 2022, at: https://www.soybeansandcorn.com/).

14 Aromatics and their derivatives

14.1 Benzene

The distillation of crude oil can yield a variety of aromatic molecules, as well as the linear and branched hydrocarbon atoms discussed in Chapters 10–13. Benzene is certainly one of the very common organic molecules that can be extracted from crude oil. While a significant amount of higher molecular weight organic molecules can be broken down to enhance the yield of liquid motor fuels, there is a significant amount and number of uses for benzene and other aromatics that they should be treated as a separate topic.

Benzene, the simplest aromatic hydrocarbon molecule derived from crude oil, can be used as a solvent for a large variety of reactions. It can be made or isolated through several different processes, but it can also be used in a variety of transformations to make a significant number of other basic commodity chemicals. Figure 14.1 and Table 14.1 give an indicator of these derived commodity chemicals.

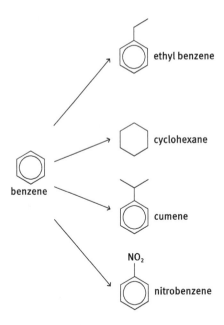

Figure 14.1: Chemicals from Benzene.

14.1.1 Catalytic reforming or steam cracking

This process produces benzene from a mixture of low molecular weight aliphatic molecules.

$$C_xH_y + H_2(g) \longrightarrow C_6H_6$$

Conditions: $PtCl_2$, 500–525 °C, >8 atm

https://doi.org/10.1515/9783110671094-014

Table 14.1: Derivatives from Benzene.

Benzene Derivative	Produces	Uses
Cumene	Phenol, acetone	Resins
Ethylbenzene	Styrene	Polymers
Cyclohexane	Adipic acid	Polymers
	Caprolactam	Polymers
Nitrobenzene	Methyl diisocyanate	Polyurethanes

The product is not pure benzene, and is referred to as reformate. Distillation is used to separate benzene from the other products [1].

14.1.2 Toluene hydrodealkylation

In this process, the methyl group in toluene is removed at elevated temperatures and pressures, using hydrogen gas.

$$H_3CC_6H_5 + H_2 \longrightarrow CH_4 + C_6H_6$$

The conditions include use of a catalyst (Pt, Cr, Mo oxide), elevated temperatures of 500–600 °C, and elevated pressure of >40 atm.

14.1.3 Toluene disproportionation

In this process, toluene is reformed to benzene and para-xylene. The latter material, discussed below, is a precursor to several nylon-based plastics, and is always in demand.

14.2 Ethylbenzene

Ethylebenzene is produced through the reaction of ethylene and benzene, as seen in Figure 14.2. Ethylene chemistry was discussed in Chapter 11. The immediate dehydrogenation reaction of this at elevated temperatures then yields styrene, which is the end product of almost all ethylbenzene production.

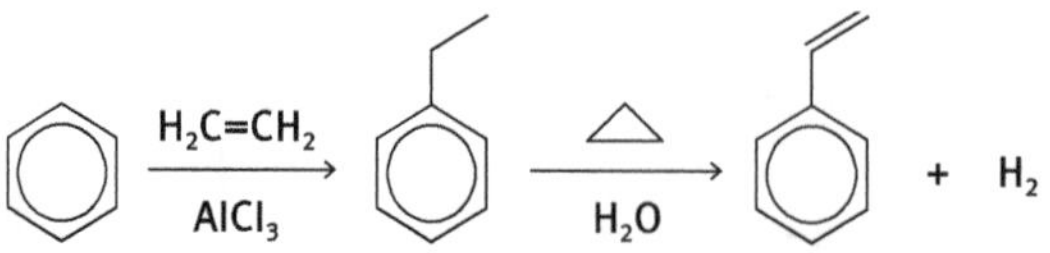

Figure 14.2: Ethylbenzene and Styrene Production.

Styrene is also co-produced with propylene oxide through what is called the POSM Process (propylene oxide styrene monomer). In this process, oxygen is added to form a peroxide, to which propylene is then added. This oxidizes the propylene to its oxide, and the alcohol is then dehydrated over alumina to yield styrene, as shown in Figure 14.3.

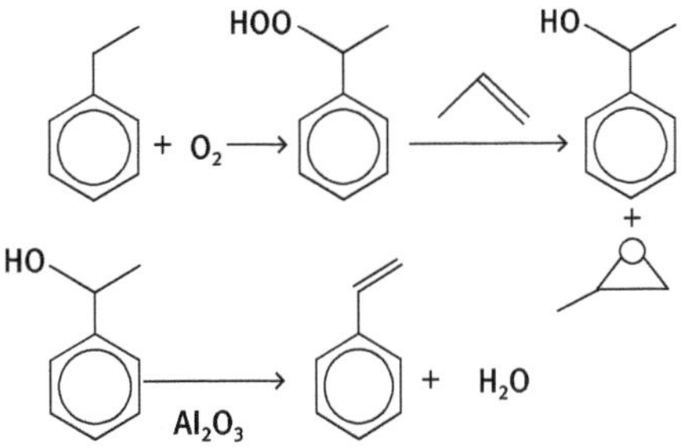

Figure 14.3: POSM Styrene Production.

14.3 Cyclohexane

Benzene can also be hydrogenated to produce cyclohexane. This requires hydrogen gas (another use for this material) and a catalyst, usually nickel or platinum.

The addition of oxygen to cyclohexane yields a mixture of cyclohexanol and cyclohexanone, which can be separated. Cyclohexanol can be ring opened using nitric acid to produce adipic acid. This straight chain bi-functional acid can in turn be used in polymer production. Cyclohexanone can be reacted to form caprolactam, which is also useful in the production of polymers, often polyamides. Production of both products is shown in Figure 14.4.

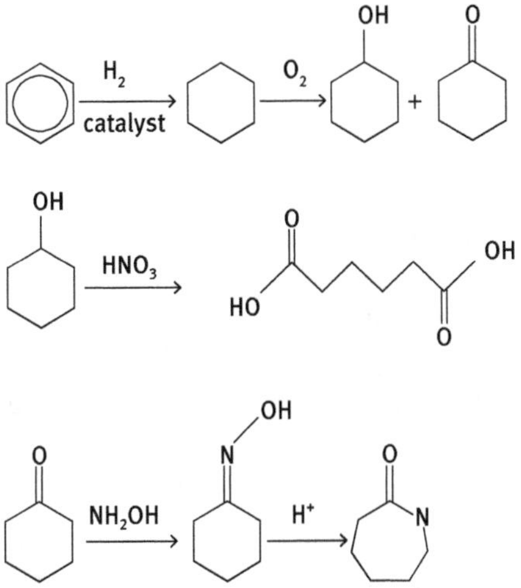

Figure 14.4: Adipic Acid and Caprolactam Production.

14.4 Nitrobenzene

A combination of nitric and sulfuric acid produces nitrobenzene. This in turn can be reduced to aniline with the addition of hydrogen. The addition of formaldehyde and then phosgene at elevated temperature directs the addition of the carbon atom of formaldehyde to the para-position of the aniline ring, and links two of them, resulting in methyl diisocyanate. This is a precursor to polyurethane based plastics, and is shown in Figure 14.5.

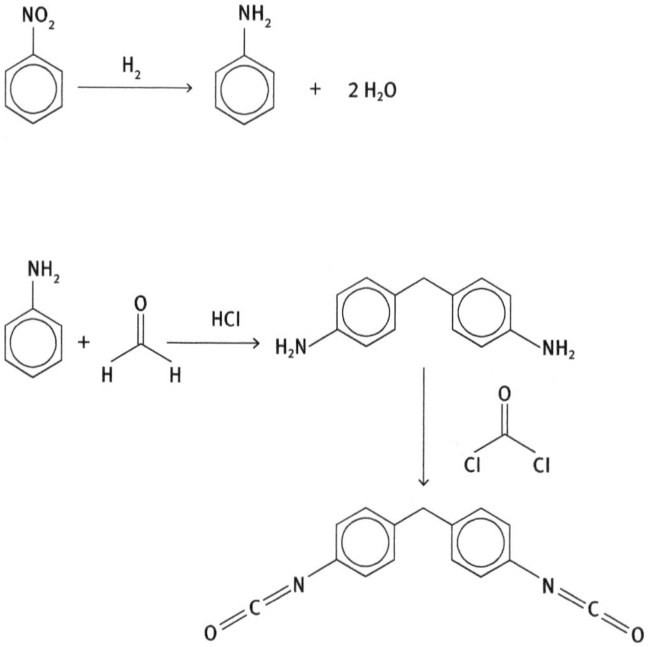

Figure 14.5: MDI Production from Nitrobenzene.

14.5 Toluene

Toluene finds use as a non-polar organic solvent, and in the past two decades has replaced some benzene use in this area. Benzene is a known carcinogen, and toluene is not known to cause the disease. Toluene is also used in a variety of paints, glues, and thinners, again because of the relative safety of use.

Toluene is also used as a gasoline additive, where it improves the performance of the fuel blend.

But toluene is also transformed into several basic commodity chemicals, as benzene is.

14.5.1 Caprolactam

Toluene can be used as a starting material for caprolactam, by first converting the toluene to benzoic acid, then hydrogenating the acid to cyclohexylcarboxylic acid. Following this, nitrosylsulfuric acid can be used to place a nitrogen adjacent to the six-membered ring. The rearrangement that follows results in caprolactam, and is shown in Figure 14.6.

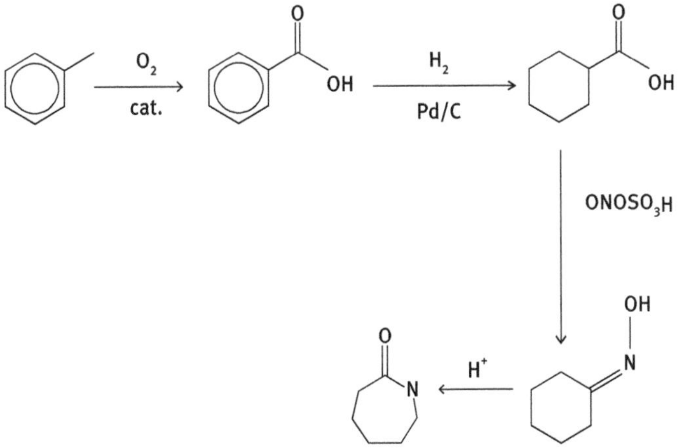

Figure 14.6: Caprolactam Production from Toluene.

14.5.2 Trinitrotoluene

The explosive, 2,4,6-trinitrotoluene (TNT) can be produced by the successive nitrations of toluene in an acidic environment. Nitric acid and sulfuric acid are used to attach two nitro groups to the toluene, in a directed fashion to the 2 and 4 positions. Further nitration at elevated temperature and pressure results in the final nitro group being attached to the now sterically crowded ring. TNT has to some extent been displaced by other explosives mixtures, but is still produced for the demolition and mining industries, and is often blended with other materials to create an array of explosive materials. This production is shown in Figure 14.7.

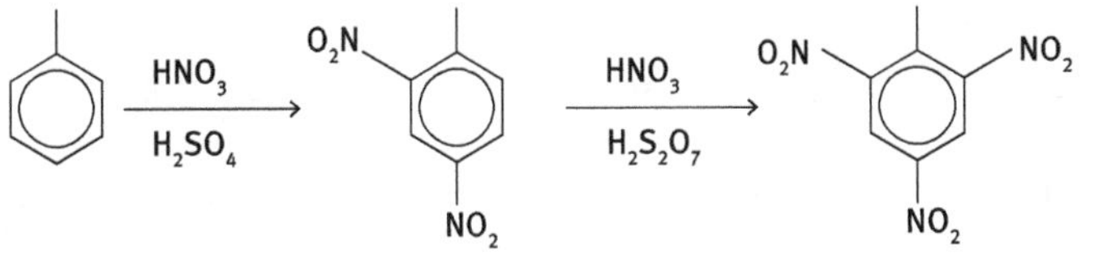

Figure 14.7: Production of 2,4,6-trinitrotoluene.

14.5.3 Toluene diisocyanate

When dinitrotoluene is reduced using hydrogen and a metal catalyst, the result is 2,4-diaminotoluene. When this is reacted with phosgene, the result is toluene diisocyanate (generally abbreviated TDI), and a by-product hydrochloric acid. TDI is then consumed in the formation of polyurethane polymers. This is an incredibly common material for polymerization, and is produced on the order of millions of tons per year. The reaction sequence is shown in Figure 14.8.

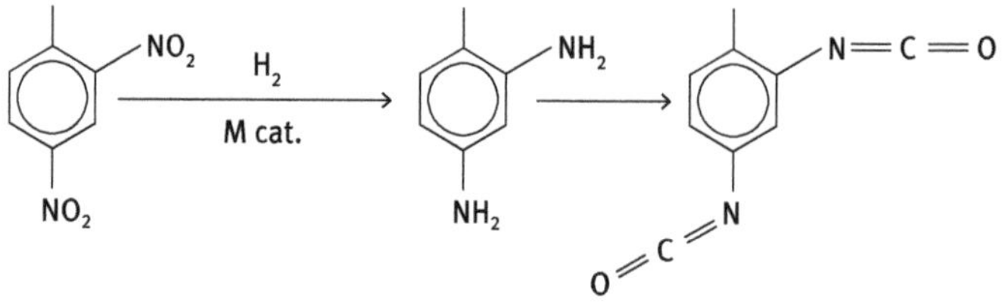

Figure 14.8: Production of TDI.

In the production of TDI, a small amount of 2,6-dinitrotoluene (roughly 5 %) is in the starting 2,4-dinitrotoluene, from the first reaction step, since the methyl group in toluene directs to the ortho- and para-positions. Upon reduction to the 2,4-diaminotoluene, this 2,6 isomer must be separated, so that the final product, the TDI, meets specifications and performance standards.

14.6 Xylene

A certain amount of xylenes, meaning the mixture of the ortho-, meta-, and para-isomers, is used as an industrial solvent when a non-polar solvent is required and there is no problem with using the mix of isomers. This is in general what is meant when the term 'xylenes' is used in the plural.

Beyond this, the major use of a specific xylene is reserved to p-xylene. At elevated temperature and pressure, and in the presence of a metal catalyst, terephthalic acid can be made in air from p-xylene. Acetic acid has been used as the solvent in most cases, and thus must be separated from the product.

The terephthalic acid product is then reacted with methanol in air, using an acidic medium to catalyze the reaction, with the final product being dimethylterephthalate, or DMT. These are then used in polymerization reactions. This reaction sequence is shown in Figure 14.9.

One use for ortho-xylene is the production of phthalate plasticizers. While this has a far smaller annual output than terephthalic acid or DMT production, it is still a large-scale process. The xylene is first reacted with oxygen at elevated temperature in the

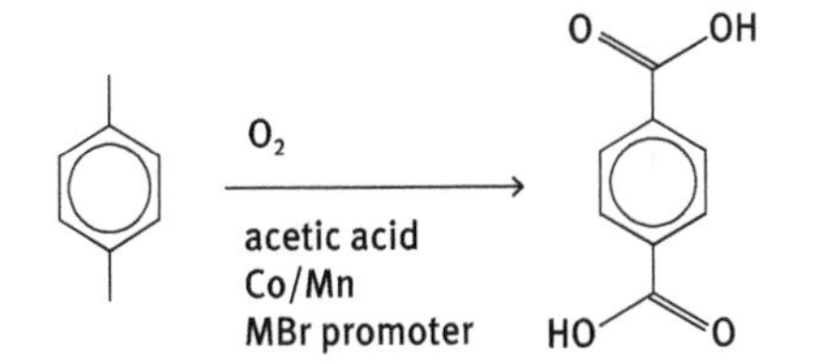

Figure 14.9: Production of Terephthalic Acid and DMT.

presence of a vanadium catalyst to produce phthalic anhydride. This anhydride can then react with long-chain alcohols to form phthalates. Several have been made, but the phthalate of industrial interest is di-(2-ethylhexyl)phthalate. The reaction sequence is shown in Figure 14.10.

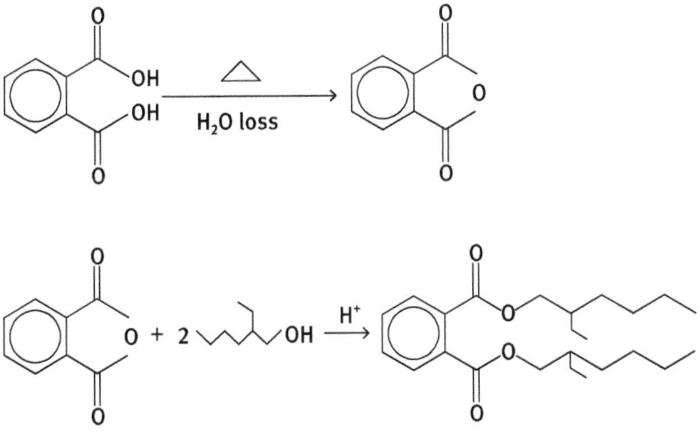

Figure 14.10: Phthalic Anhydride and Phthalate Production.

Such materials are used as plasticizers in various plastics that would otherwise be rigid. There continues to be some controversy over the end-use products into which such plasticizers are incorporated. They can leach from the material, and be ingested, and may be carcinogenic.

14.7 Pollution and recycling

Almost all the basic commodity chemicals discussed here are used in producing some further downstream material, and/or end-use consumer product. Thus, recycling becomes a matter of plastics recycling, which will be discussed in Chapter 15. When benzene, toluene, and xylene are used as solvents, recycling may occur within an operation, if it is economically feasible. If it is not, spent solvents sometimes find a second use as feeds in power plants, generating energy through their combustion.

Bibliography

[1] ICIS. Website. (Accessed 2 December 2022, at www.icis.com/).

15 Polymers

15.1 Introduction

We have already alluded to how profoundly the large scale production of plastics has changed our world, its peoples, and our quality of life. Modern medicine, high speed transportation, and protective gear for sports are three disparate examples of how areas have been influenced by plastics to enhance the quality of human life. Almost all cultures have a bronze age followed by an iron age. The reason iron follows bronze is essentially that the melting point of iron is the higher of the two, and thus someone in the culture must find or develop a fuel that can melt iron. Globally, we may now be living in a "plastics age," as these materials have been so completely integrated into virtually all human cultures. For a plastics age to exist, not only did crude oil need to be discovered, but methods of distilling it into pure materials had to be developed, and processes of polymerizing those resulting small molecules had to be developed and brought up to large scale use. While the materials we will examine here have been discussed in previous chapters, we will examine the chemistry of the plastics in more detail.

15.2 Resin identification codes, RIC 1–6

Resin Identification Codes have been developed for six plastics that have the majority of consumer uses. There are certainly many more plastics in large scale use than six, but these six form a very large portion of the whole gamut of plastics that exist, and are recyclable in some way. Several organizations devoted to plastics uses, recycling, and awareness of the lifecycle of these materials exist [1, 2, 3, 4, 5, 6, 7, 8].

The six major plastics are listed by RIC in Table 15.1.

Table 15.1: Names and RICs of Common Polymers.

RIC	Name	Common Abbreviation	Example Uses
1	Polyethylene terephthalate	PETE, PET	Beverage bottles
2	High density polyethylene	HDPE	Plastic lumber
3	Polyvinylchloride	PVC	Piping
4	Low density polyethylene	LDPE	Plastic bags
5	Polypropylene	PP	Food containers, living hinges
6	Polystyrene	PS	Insulating material, cups

In the United States of America, the RIC "7" is reserved for other plastic, blended plastics, or plastics that have already been recycled once. The reason these six polymers have such designations is simply that they are produced annually on such a large scale.

https://doi.org/10.1515/9783110671094-015

The Society of the Plastics Industry developed the current RIC system in 1988, with the aim being ease of recycling materials through a pre-recycling separation, although the system now appears to be followed internationally [8, 1].

15.3 Thermoplastics

Thermoplastics are polymer materials that become soft above a certain temperature, and hard or glass-like below that temperature. This temperature, called the glass transition temperature and designated by the symbol T_g, is different for each material. While there are many plastics that have a specific T_g, some only melt (this temperature is usually designated by T_m for melting temperature). The more common plastics are listed in Table 15.2. There are many others, often occupying some niche in terms of a manufactured end use item.

Table 15.2: T_g or T_m of the Common Polymers and Blends.

RIC	Name	Abbrev.	T_g (°C)	T_m (°C)
1	Polyethylene terephthalate	PETE	75	255
3	Polyvinyl chloride	PVC	80	
5	Polypropylene	PP		160
6	Polystyrene	PS		240
	Acrylonitrile butadiene styrene	ABS	105	
	Styrene – acrylonitrile	SAN	115	

15.4 Thermosets

The term "thermoset" simply means a material that irreversibly forms the end product, usually after heat or pressure is applied. As with thermoplastics, there are numerous thermosets that have found some use in a specific industry. Examples include those shown in Table 15.3.

There exist a variety of methods for molding or forming thermosets. They include:

- Compression molding. This is used extensively, and requires the use of increased pressure, as the name implies. It forces the material into the desired shape, often of some end use product.
- Extrusion molding. Long objects, such as piping, are easily produced by forcing the warm thermoset through an extruder. Different extruder heads can be used to create specific shapes.
- Injection molding. This process involves pushing the warm, elastic material into a mold – injecting it.

Table 15.3: Thermoset Examples.

Thermoset	Example Uses
Bakelite	Consumer end use items
Duroplast	Fiber reinforced resin
Melamine resin	Consumer end use objects, e. g.: cookware
Epoxy resin	Adhesives
Polycyanurate	Electronics
Polyester fiberglass	Insulating materials
Polyimide	Electronics
Polyurethane	Coatings, adhesives, insulating foam
Urea-formaldehyde resin	Particle board and wood adhesive
Vulcanized rubber	Tires, clothing fiber (see Chapter 21)

–   Spin casting. This technique uses centrifugal force to push warm or molten plastic into a desired shape, then allows it to cool, retaining the shape.

Precisely because thermosets are formed irreversibly, it is difficult to recycle them. Secondary uses are sometimes found, but usually involve some sort of mechanical degradation of the thermoset material.

15.5 Polyethylene terephthalate (PETE)

Polyethylene terephthalate is a thermoplastic polymer that is composed of ethylene glycol and terephthalic acid, both of which were discussed in previous chapters. The reaction to form this material, shown in Figure 15.1, is classed as a condensation polymerization because water is produced as a by-product. The reaction occurs at elevated temperature ($>220\,°C$), and the co-product water is distilled off.

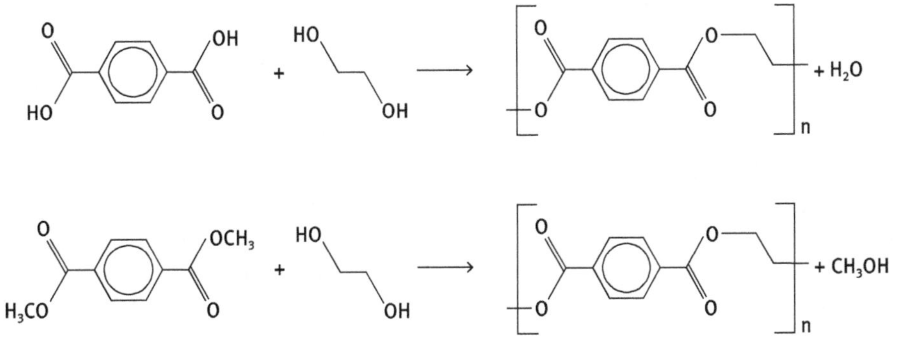

Figure 15.1: PETE Formation.

As can be seen from Figure 15.1, PETE can also be produced using dimethyl terephthalate as a starting monomer. This occurs at 150–200 °C in the presence of a base catalyst. The resultant polymer co-produces methanol instead of water in this version of the reaction. The methanol can be recovered by distillation.

The annual tonnage for PETE production remains in the millions, with the majority of the product being consumed in synthetic fibers, and ultimately clothing and other synthetic cloth uses (such as sails and tarps). The term "polyester," when applied to clothing, generally means materials made from this plastic. The more familiar use, the production of plastic bottles, accounts for a smaller portion of the whole.

15.6 Low density polyethylene and high density polyethylene (LDPE and HDPE)

These two polyethylenes will be treated together because they come from the same monomer, even though their polymerizations produce very different final products, when compared in terms of physical properties. In a simplified schematic, as shown in Figure 15.2, ethylene is simply mono-polymerized, yielding long chains of $-CH_2-$ repeat units. The production of all kinds of polyethylene has become a billions of dollars per year industry, and the various densities of polyethylene each assume a certain amount of market share. In general, the polymerization of ethylene (and propylene, treated below) can be said to be done via a Ziegler–Natta Process, meaning through the use of a titanium catalyst and aluminum co-catalyst. Some delineate further, and state that ethylene polymerization is through Ziegler catalysis, reserving the term 'Ziegler–Natta' for propylene polymerizations.

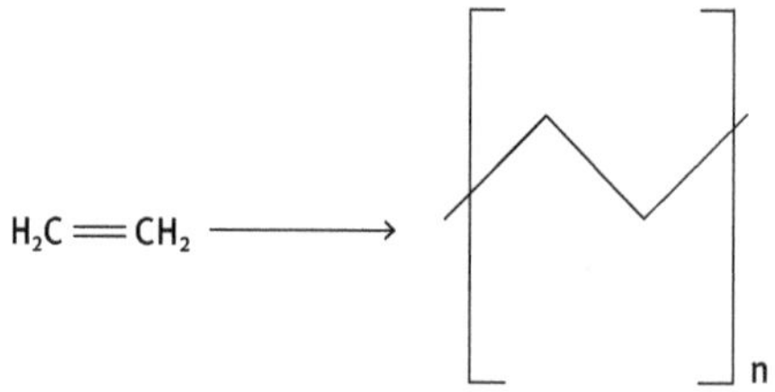

Figure 15.2: Polyethylene from Ethylene.

Low density polyethylene (LDPE) depends upon a free radical mechanism. Various initiators can be used, all of which can be classified as stable free radicals. The transfer of a free radical to the main, growing polymer chain produces free radicals on the growing chain, which allows for greater branching possibilities. Increased numbers of branches on the polymer chains results in a lower density material than if there were few or no branches.

There is a wide array of end use products that are made of LDPE, including packaging materials that require some flexibility, such as jar and can lids, as well as 6-pack holders, and flexible, deformable packages and containers.

High density polyethylene (HDPE) has fewer branches in its overall structure, and thus a greater material density. It is also a more crystalline structure than LDPE. Its polymerization is usually performed in the presence of a metal oxide catalyst, but not in all cases.

But HDPE is used in many applications precisely because it has a high strength to density ratio. For this reason, it finds uses in a variety of piping, both in residential and municipal settings, in plastic furniture, and in plastic lumber.

15.7 Polypropylene (PP)

As mentioned above, polypropylene (PP) is manufactured through Ziegler–Natta catalysis, although metal oxide catalysis also works. It is another thermoplastic material, but one with an elevated melting point. Because of this relatively high melting point (roughly 170 °C), PP has found extensive use in medical settings where materials must be repeatedly sterilized. The chemistry of PP formation looks like that of polyethylene formation, in that it is the double bond that is opening and forming new bonds. Figure 15.3 shows a simplified reaction.

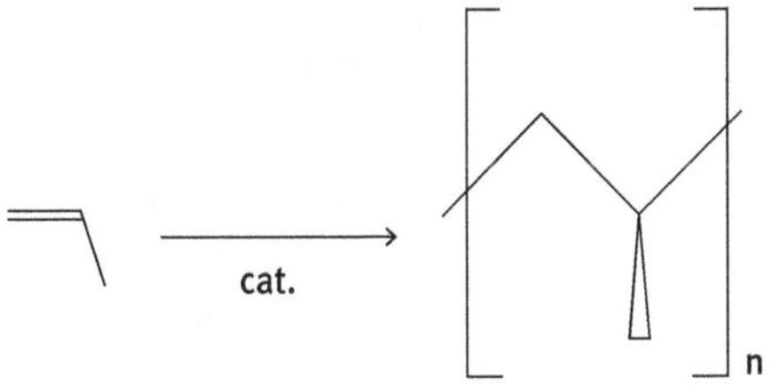

Figure 15.3: Polypropylene Formation.

Unlike any form of polyethylene, polypropylene exists in three possible forms, based on the position of the methyl groups pendant to the main chain. When the repeat unit is always positioned on the same side of the chain (front or back, depending on one's view), the material is referred to as being isotactic. In the configuration in which the methyl groups form a pattern of back and forth for every two repeat units, the term is syndiotactic polypropylene. When there is no order to the placement of the methyl groups, the material is atactic polypropylene. Figure 15.4 shows the repeat unit structure of isotactic and syndiotactic polypropylene.

Like polyethylene, polypropylene is manufactured in the range of tens of millions of tons annually, with much of the material making it to end use consumer products. The term 'living hinge' is used in connection with polypropylene because this material is able to be flexed repeatedly. The hinges on food products that snap open and closed,

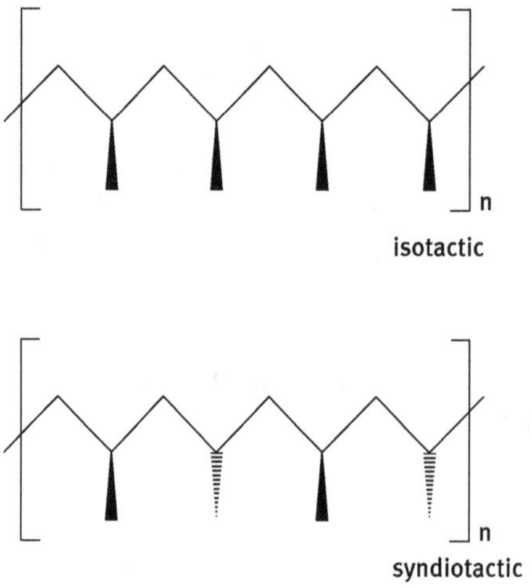

isotactic

syndiotactic

Figure 15.4: Isotactic and Syndiotactic Polypropylene.

but never disconnect from the product, are living hinges. As mentioned, many objects used in medical settings are made of PP, as are plastic furniture and lumber, and as is plastic sheathing for electrical cabling.

15.8 Polyvinyl chloride (PVC)

The polymerization of vinyl chloride to produce polyvinylchloride (PVC) occurs much like the polymerization of polypropylene. It has a slightly lower melting point than PP (140 °C), but finds use in piping and other material where flexibility matters, because plasticizers (such as the phthalates discussed in Chapter 14) can be added to it during the formation of end use products. The physical characteristics of PVC, and the ability to change them with the addition of various plasticizers, make it a useful material in numerous long-term, end use applications.

15.9 Polystyrene (PS)

Similarly to PP and PVC, and as mentioned in Chapter 13, there are three possibilities for the tacticity of polystyrene, depending on the positioning of the pendant phenyl rings, but only the atactic version that has found widespread commercial use. When air is blown through polystyrene during the formation of objects, the very common Styrofoam is the result. The name Styrofoam is trademarked by Dow Chemical.

The uses of Styrofoam are often intertwined with any object that in some way involves heat transfer, or the prevention of it. Cups, containers, and coolers are often made of Styrofoam, or with it in some structural or insulating capacity. The material is more than 95 % air, and can be molded into virtually any shape.

15.10 Pollution, recycling possibilities, and by-product uses

Few chemical commodities stir up more passion and debate than the use and recycling of plastics. Plastics recycling programs are sometimes dictated by economics, in which one state or area has store owners pay a small amount back for each item returned, or by the perception of environmental friendliness, in which drop-off points are established for plastics to be returned. Plastics are demonized when they end up in the waste stream, because their decomposition rates are extraordinarily slow, or because they end up concentrating in specific areas (the area in the Pacific Ocean now sometimes referred to as the "Pacific garbage patch" may be the world's largest example).

Yet in theory, plastics should be endlessly recyclable. Especially noteworthy in this regard are thermoplastics, which can be re-melted and reformed.

15.10.1 PETE

Recycling of plastic beverage bottles has become established in many developed countries. Such bottles are almost always made from PETE, but because they are often wrapped with another plastic which has the logo and coloring of the product on them, it becomes difficult to recycle such plastic to another end use product – another bottle – of the same quality. Thus, what is sometimes termed "down cycling" occurs, where the recycled plastic is used for end use products such as furniture or rugs. This lengthens the amount of time a material is used, but ultimately does not keep it from being discarded. The furniture and rugs just mentioned can be further used, but the color tends to devolve to an ugly green-brown, suitable only for use as dunnage in containers, or other uses that remain out of sight.

15.10.2 Polyethylene, polypropylene, and polyvinyl chloride

These materials are assigned RICs, but are not recycled to the extent that PETE is. In large part, this is simply because more PETE is used in the production of consumer packaging that can be discarded after use, such as beverage bottles. Also, the resin quality of PETE used in beverage bottles and containers tends to be of higher quality – it is often clearer – than PE, PP, or PVC that is being considered for recycling.

15.10.3 Polystyrene

Because of its extremely low density, and because most of it is air, Styrofoam is not recycled to any large extent. The cost of transportation of the spent material is always prohibitive, since the Styrofoam has such low density. The Dow Company does recycle this material internally, but no post-consumer Styrofoam recycling programs currently exist [3].

Bibliography

[1] Chemistry Industry Association of Canada. Website. (Accessed 2 December 2022, at: https://canadianchemistry.ca/).
[2] China Plastics Processing Industry Association. Website. (Accessed 2 December, 2022, http://www.cppiaeps.com/).
[3] Dow Chemical Company. Website. (Accessed 2 December 2022, at www.dow.com).
[4] Japan Plastics Industry Federation. Website. (Accessed 2 December, 2022, at http://www.jpif.gr.jp/english/).
[5] Plastics Industry Manufacturers of Australia. Website. (Accessed 2 December, 2022, https://www.pima-org.net/).
[6] Plastics New Zealand. Website. (Accessed 2 December 2022, at www.plastics.org.nz/).
[7] Plastics Europe. Website. (Accessed 2 December 2022, at www.plasticseurope.org/).
[8] Society of the Plastics Industry. Website. (Accessed 2 December, 2022, at www.plasticsindustry.org/).

16 Coatings and adhesives

16.1 Introduction

Coatings and adhesives have become a prevalent part of the chemical industry, and have found uses in or as literally tens of thousands of consumer end products. Our discussion here will also include pigments, because these are often used as coatings. Simply put, many of them are paints, and thus coat a great many objects and items.

16.2 Coating types

A coating is a thin layer of any chemical material applied to another material or object, often referred to as a substrate. The coating is applied to the main object or substrate for one or more reasons, such as improving and lengthening the lifetime of the substrate, enhancing or optimizing its properties, changing the appearance of the substrate, or chemically altering the substrate. When coatings are paints and varnishes, the twin end results are usually a better looking product, and one that is protected from its environment.

Printing can be considered a type of coating, especially when we consider the printing of images as well as lettering. While the term 'printing' invokes images of black letters on white paper, there is a great deal more to it, including images and lettering on placards, large signs, and billboards.

Coatings are usually applied as liquids, but these can be aerosolized during application, or can even be in solid form, as a powder. While powders are often mixed with some solid, this is not always the case. In some cases, they can be mixed just as easily with liquids.

16.3 Adhesives and binders

As the name implies, adhesives and binders are chemicals that bind one object or material to another. There are numerous ways to sub-divide adhesives, including natural and synthetic, as well as single-component and double-component. Numerous adhesives involve polyurethanes, which we have touched upon in previous chapters.

One part adhesives, as the name implies, require no mixing before the adhesive is applied. Two-part, or multipart, adhesives require a mixing of the two components just prior to applying the adhesive to the surfaces to be bonded.

Some adhesives are reactive enough that they are stored in some container pre-mixed with a volatile organic solvent. The familiar smell of model glue is actually the toluene solvent in the glue. After applying it, the toluene evaporates, leaving the two pieces that have been joined permanently together.

https://doi.org/10.1515/9783110671094-016

16.4 Fillers

Many composite materials and plastics have fillers added to them. The only real purposes of a filler are to lower the amount of a more expensive material required in the final product, or to in some way change the properties of the end material in a positive way. In the best case scenario, one filler can do both, thus making a superior product, and doing so at a lower cost to the producer.

The list of fillers can become long, but common ones include:
–   Carbonates
–   Ground Calcium Carbonate (GCC)
–   Precipitated Calcium Carbonate (PCC)
–   Dolomite
–   Barium sulfates
–   Silicates
–   Feldspar
–   Mica
–   Nepheline syenite
–   Talc
–   Clay (i. a. kaolin)
–   Wollastonite
–   Quartz
–   Diatomite
–   Synthetic silicon dioxide
–   Hydroxides
–   Aluminum trihydroxide (ATH)
–   Magnesium hydroxide (MDH)
–   Graphite
–   Carbon black
–   Fiberglass and glass spheres

This list is taken from a recently published Ceresana marketing report [1], and includes several materials we discuss in various chapters of this book.

There are many others in addition, but they are either less common than those listed, or they are used in more specific, niche areas.

Among fillers, calcium carbonate (the subject of Chapter 5) is often listed first because it is so widely and heavily used. It can be added to cement mixtures, used as a paint whitener, and added to the absorbent of infant disposable diapers, to name just a few very different examples. It finds use as more than a filler however; calcium carbonate can be ingested by patients who require more calcium in their diets.

Additionally, graphite and carbon black make this list, and are discussed in Chapter 7. The market for carbon black as both a filler and a pigment remains a large one, with carbon black finding uses in materials as widely different as paints and vehicle tires.

16.5 Pigments

Pigments can be divided in numerous ways, with inorganic and organic being a rather simple one.

Inorganic pigments are materials derived from an inorganic source that are added to any material or blend of materials to produce a specific color. While the following list is large, it cannot represent a complete list, as chemists and chemical engineers con-

Table 16.1: Inorganic Pigments.

Color	Name	Formula	Example Use
Red	Cadmium red	$CdSe$	Artist paints
	Red ochre	Fe_2O_3	Paints
	Burnt sienna	Fe_xO_y	Paints
	Red lead	Pb_3O_4	Primer paint
	Vermillion	HgS	Oil paints
Orange	Cadmium orange	$CdSSe$	Artist paints
	Chrome orange	$PbCrO_4/PbO$	Minimal use today, paints
Yellow	Orpiment	As_2S_3	Paints
	Cadmium yellow	CdS	Artist paints
	Chrome yellow	$PbCrO_4$	Paints
	Cobalt Yellow	$K_3Co(NO_2)_6$	Oil-based paint
	Yellow ochre	$Fe_2O_3 \cdot 6H_2O$	Oil-based paint
	Stannic sulfide	SnS_2	Coatings
Green	Cadmium green	Cr_2O_3	Artist paints
	Paris green	$Cu(C_2H_3O_2)_2 \cdot 3Cu(AsO_2)_2$	Paints, fireworks
	Schloss green	$CuHSO_3$	Minimal use, paints
	Viridian	$Cr_2O_3 \cdot xH_2O$	Paints
Blue	Cobalt blue	$CoSnO_2$	Ceramics and paints
	Egyptian blue	$CaCuSi_4O_{10}$	Paints, security ink
	Han blue	$BaCuSi4O_{10}$	Ceramics
	Prussian blue	$Fe_7(CN)_{18}$	Paints, blueprints
	Ultramarine	$Na_8Al_6Si_6O_{24}S_2$	Inks, paints
Purple	Cobalt violet	$Co_3(PO_4)_2$	Paints
	Han purple	$BaCuSi_2O_6$	Paints
	Manganese violet	$Mn(NH_4)PO_4$	Ceramics
Brown	Sienna	Fe_xO_y	Paints
	Raw umber	$Fe_2O_3/MnO_2/Al_2O_3$	Ceramics
Black	Carbon black	C_6	Tires
	Ivory black	C_x	Printing ink
	Lamp black	C_6	Inks and paints
White	Titanium white	TiO_2	Food
	Antimony white	Sb_2O_3	Ceramics and glass
	Barium sulfate	$BaSO_4$	Paper and paint
	White lead	$(PbCO_3)_2 \cdot Pb(OH)_2$	Oil paints
	Zinc white	ZnO	Paper coating, paint

tinue to look for new materials that can produce desired colors in any material that will eventually be used by consumers.

Since there is no overarching theory concerning pigments, the following list of pigments in Table 16.1 is arranged simply using the time-worn acronym ROY G. BIV, for the color spectrum.

Inorganic chemistry is not the sole domain from which pigments are derived, although it has traditionally been one from which people have derived paints with bright colors. There are certainly organic pigments as well. Indeed, a trade organization for these latter exists in large part to ensure there is monitoring of any toxicity from pigments added to consumer products [3]. Specifically, there is always concern that colorants and dyes used in clothing and textiles, as well as leather goods or what are called "food contact materials," may have adverse effects on the immediate environment, or on the food packaged by a dyed or colored material. Some of the more common organic dyes are shown in Table 16.2, again, simply listed according to the color spectrum.

Table 16.2: Organic Dyes and Pigments.

Color	Name	Use
Red	Alizarin	Textile dye
	Cochineal red	Food coloring
	Lycopene	Food coloring
	Paprika/henna	Hair dye
	Rose madder	Paint, textile dye
Orange	Annatto	Food coloring
Yellow	Gamboge	Textile dye
	Indian yellow (euxanthin)	Oil paints
Green	Chlorophylin	Food coloring
Blue	Indigo	Paint, textile dye
Purple	Tyrian purple	Textile dye
	Mauveine	Textile dye

The Color Pigments Manufacturers Association is another organization devoted to pigments and their chemistry and manufacture, which appears to focus much more on inorganic materials [2].

16.5.1 Recycling

Most adhesives, binders, fillers, and pigments are not recycled, although some are chosen for a commercial use specifically because they do not interfere with another product that is recyclable and thus can be recycled themselves. The example of adhesives used to close and seal cardboard or paperboard boxes is one in which a significant tonnage of adhesive is used annually, that is later recycled along with the paper product. The paper industry is discussed in more detail in Chapter 18.

Bibliography

[1] Ceresana. Website. (Accessed 6 December 2022, at www.ceresana.com).
[2] Color Pigments Manufacturers Association, Inc. Website. (Accessed 6 December 2022, at www. pigments.org).
[3] The Ecological and Toxicological Association of Dyes and Organic Pigments Manufacturers, ETAD. Website. (Accessed 6 December 2022, at www.etad.com).

17 Fertilizers and pesticides

17.1 Introduction

Fertilizers have an ancient history, in that mankind has known how to aid in the growth of crops for millenia. But only in the past two centuries has the production of fertilizer become an industrial – scale process, and not the re-use of untreated materials, such as animal wastes, to enhance plant growth. We saw in Chapter 4 how nitrogen-based fertilizers are produced on an industrial scale. Here, we will examine the use of them and phosphorus-containing fertilizers, as well as materials that add potassium to the soil. This is a large enough enterprise, and an important enough one, that there are several national and international organizations that are dedicated to proper use of fertilizers [1, 2, 3, 4, 5, 6, 7, 8].

Additionally, we will discuss pest control – which is often referred to as crop protection – since insects and other vectors which consume crops in competition with mankind have also been with us since the beginning of agriculture.

17.2 Nitrogen-containing materials

The major nitrogen-containing fertilizers were discussed in Chapter 4. The direct injection of ammonia into soil, or the use of ammonia-derived fertilizers such as ammonium nitrate, remains one of the largest consumer end-use businesses in the developed world.

17.3 Phosphorus-containing materials

While several phosphorus – containing materials can be used in fertilizers, dipotassium phosphate is a highly water soluble commodity that is often used because it delivers a high amount of phosphorus to the growing plant. Additionally, K_2HPO_4 is used as a food additive, but different grades are required for food versus fertilizer use.

The reaction chemistry for the production of dipotassium phosphate appears straightforward, as a condensation using potassium hydroxide, as shown below.

$$H_3PO_4 + 2KOH \longrightarrow K_2HPO_4 + 2H_2O$$

The phosphoric acid is in turn manufactured from phosphate rock, which is found in several places throughout the world. In the United States, Florida and North Carolina account for most phosphate rock production, a significant portion of which is used for fertilizer. Figure 17.1 shows production sources as a percentage, with the total being 220 million metric tons (based on production figures from 2021).

https://doi.org/10.1515/9783110671094-017

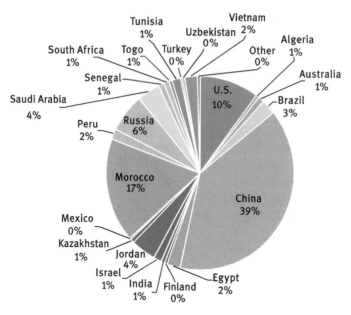

Figure 17.1: Phosphate Rock Production.

17.4 Potassium containing materials

Potash is used to manufacture potassium chloride for use in fertilizers. According to the US Geological Survey, approximately 85 % of mined potash is in some way used for fertilizer [9]. Curiously, according to the USGS and most organizations devoted to fertilizers [5, 6, 7, 4, 9], potash is often reported as K_2O equivalent, even though this compound is not what is routinely delivered to the crop. Potash is a mixture of potassium-containing compounds, all of which are very water soluble because of the presence of the potassium. Indeed, mining of potash can be through solution wells, in addition to traditional mining, because injecting hot water into a potash formation will dissolve much of it, making it easy to extract. But potassium oxide is too basic to be stable in water. What normally ends up being delivered to growing crops is KCl.

17.5 Mixed fertilizers

Mixed fertilizers tend to deliver nitrogen, phosphorus, and potassium to the soil all at the same time. In the United States, a three number system has been developed that indicates the percentage of these three elements in the mix, by weight. Thus, the well-known 10-6-4 fertilizer contains 10 % nitrogen, 6 % phosphorus, and 4 % potassium.

17.5.1 Production of mixed fertilizers

We have already seen that ammonia can be made by the direct combination of the elements at elevated pressure and temperature in the presence of a catalyst. Urea, made from ammonia and CO_2, is a more convenient way to blend the nitrogen component into a mixed fertilizer since it is a white, crystalline solid, and thus is often used. Potassium is introduced in the form of potassium chloride, and phosphorus is in the form of phosphoric acid.

Mixed fertilizers can be delivered as time delayed, or delayed release fertilizers as well. In such cases, the components are mixed with a polymer resin, such as urea-formaldehyde, so that they dissolve more slowly in water, and thus enter the soil over a longer period of time. Certain sulfur-coated fertilizers – meaning the fertilizer particles or pellets are encased in sulfur – are also used in this manner, as the addition of small amounts of sulfur to some soils is beneficial.

17.6 Pesticides

As mentioned in the introduction, what are called "pests" have been present since the very beginning of agriculture. Imagine the grief of early farmers when they, after all the hard work of ploughing, sowing, and growing a crop, watched helplessly as some insect swarm came and ate everything in their fields. Crops could simply vanish in a cloud of insects, leaving the people with nothing but a long, hungry stretch of time before them. It is little surprise then that the advent of pesticides was greeted by farmers as a miracle.

In the modern day, the term pesticide has acquired a much more sinister undertone, as the mis-application or over application of these materials has caused serious environmental damage and human health problems. There is no denying that large scale, wide spread spraying of some herbicides and insecticides has done serious damage, but it should be noted that another method has never been found to provide food for the more than seven billion people currently on Earth.

17.6.1 Herbicides

The definition of a weed is simply a plant that is growing in a place where humans do not want it to. This makes the concept of a herbicide a difficult one to bring to reality, as any material that will kill one plant will most likely kill another. Nevertheless, since the Second World War, there has been significant research into the development of selective herbicides. Mostly, herbicides are used as crop protection chemicals, to increase crop yields. But military forces have at times been interested in the use of non-selective herbicides, so that an entire area might be defoliated. Perhaps the most notable example of this is Agent Orange, used by United States forces in the Vietnam War. The material was an effective defoliant, but caused significant and severe health problems to military personnel and native peoples who were exposed to it.

While there are many herbicides, a partial list, including the aim of each, is shown in Table 17.1.

Table 17.1: Herbicides.

Trade Name	Chemical Name	Formula	Intended Use
Selective herbicides			
Aminopyralid	4-amino-3,6-dichloropyridine-2-carboxylic acid	$C_6H_4O_2N_2Cl_2$	Broadleaf, thistles, clover
Atrazine	1-chloro-3-ethylamino-5-isopropylamino-2,4,6-triazine	$C_8H_{12}N_5Cl$	Grasses and broadleaf
Citrus oil	Limonene	$C_{10}H_{16}$	Broadleaf
Clopyralid	3,6-dichloro-2-pyridine carboxylic acid	$C_6H_3O_2NCl_2$	Broadleaf
Dicamba	3,6-dichloro-2-methoxybenzoic acid	$C_8H_6O_3Cl_2$	Broadleaf
Fluroxypyr	[(4-amino-3,5-dichloro-6-fluoro-2-pyridinyl)oxy]acetic acid	$C_7H_5O_3N_2Cl_2F$	Broadleaf
Pendimethalin	3,4-dimethyl-2,6-dinitro-N-pentan-3-yl-aniline	$C_{13}H_{19}O_4N_3$	Grasses and broadleaf
Picloram	4-amino-3,5,6-trichloro-2-pyridine carboxylic acid	$C_6H_3O_2N_2Cl_3$	Trees
Vinegar	Acetic acid	C_2H4O_2	Broadleaf
Non-selective herbicides			
2,4-D	(2,4-dichlorophenoxy)acetic acid	$C_8H_6O_3Cl_2$	
2,4,5-T	2,4,5-trichlorophenoxyacetic acid	$C_8H_5O_3Cl_3$	
Glyphosate	N-(phosphonomethyl)glycine	$C_3H_5O_5NP$	
Imazapyr	(R,S)-2-(4-methyl-5-oxo-4-propan-2-yl-1H-imidazol-2-yl)pyridine-3-carboxylic acid	$C_{13}H_{15}O_3N_3$	
Paraquat	1,1'-dimethyl-4,4'-bipyridinium dichloride	$C_{12}H_{14}N_2Cl_2$	
Phosphinothricin	Ammonium (2-amino-4-(methyl-phosphinato)butanoate)	$C_5H_{15}O_4N_2P$	
Sodium chlorate		$NaClO_3$	

17.6.2 Insecticides

As with herbicides, there are a wide variety of insecticides that have been developed in the past seventy years. Most have been produced using traditional organic syntheses. Some have been used on crops, but developed for other reasons. One example is the well-known DDT, production of which has now been phased out, and which is now associated with chemically-induced disease. This was first pursued not because of the perception or hope of increased crop yields, but rather in the pursuit of an insecticide that would protect United States Marines, soldiers, and sailors who were fighting in the Pacific Theater in the Second World War. The use of this material brought the number of deaths and incapacitation from insect borne illnesses down significantly. This was the first time that a military employing soldiers unaccustomed to life in tropical lands, with all the attendant insects of such a climate, saw more fatalities from the enemy than from regional diseases.

A listing of present and past insecticides would be very large. Thus, a partial listing of insecticides that tend to be better known is shown in Table 17.2.

Table 17.2: Insecticides.

Trade Name	Chemical Name	Formula	Intended Use
Acephate	N-(methoxy-methylsulfanylphos-phoryl)acetamide	$C_4H_9NO_3PS$	Aphids
Aldrin	1,2,3,4,10,10-hexachloro-1,4,4a,5,8,8a-hexahydro-1,4,5,8-dimethanonaphthalene	$C_{12}H_8Cl_6$	(phased out)
Cinnamaldehyde	(2E)-3-phenylprop-2-enal	C_9H_8O	Fungicide
DDT	1,1,1,-trichloro-2,2-di(4-chlorophenyl)ethane	$C_{14}H_9Cl_5$	Mosquitoes
Diazinon	O,O-diethyl-O-[4-methyl-6-(propan-2-yl)pyrimidin-2-yl] phosphorothioate	$C_{12}H_{21}O_3N_2PS$	Fleas, roaches
Endrin	(1aR,2S,@aS,3S,6R,6aR,7R,7aS)-3,4,6,9,9-hexachloro-1a,2,2a,3,6,6a,7,7a-octahydro-2,7:3,6-dimethanonaphtho[2,3b]oxirene	$C_{12}H_8OCl_6$	(phased out, banned)
Heptachlor	1,4,5,6,7,8,8-heptachloro-3a,4,7,7a-tetra-hydro-4,7-methano-1H-indene	$C_{10}H_4Cl_7$	Multi-species
Malathion	Diethyl 2-[(dimethoxyphosphoro-thioyl)sulfanyl]butanedioate	$C_{10}H_{19}O_6PS_2$	Mosquitoes
Methomyl	(E,Z)-methyl N-{[(methylamino)car-bonyl]oxy}ethanimidothioate	$C_5H_{10}O_2N_2S$	Flying insects
Methylbromide	Bromomethane	CH_3Br	Fumigant for strawberries
Nicotine	3-[(2S)-1-methylpyrrolidin-2-yl]pyridine	$C_{10}H_{14}N$	(limited use)
Parathion	O,O-diethyl O-(4-nitrophenyl) phosphorothioate	$C_{10}H_{14}NO_5PS$	Ticks, mites
Thymol	2-isopropyl-5-methylphenol	$C_{10}H_{14}O$	Multi-species

As can be seen from Table 17.2, some insecticides have been banned or severely restricted because they are highly poisonous to other forms of life besides insects. A continuing problem with the formulation of insecticides is that insects have extremely short life-spans, and thus produce new generations of insects faster than other faunal forms of life. This translates to a quicker ability to adapt to threats, including chemical threats, to their species. Therefore, one insecticide may not be effective against insects for decades. However, higher animals such as birds which eat insects, do not live, die, and adapt as quickly as the insects. Thus, insecticides tend to be more poisonous to birds and other higher life forms.

17.7 Pollution and recycling

Virtually no fertilizers or pesticides are recycled, because all are used and disseminated in growing crops and in crop protection. The environmental fate of both, but especially of pesticides, is one reason the Environmental Protection Agency dedicates considerable resources to their monitoring [2].

Unfortunately, the mis-application of both fertilizers and pesticides has caused significant pollution problems in almost all areas where crops are grown. In the center of the United States, the non-point source pollution from both fertilizers and pesticides tends to run into waterways that ultimately run to the Mississippi River. This in turn empties into the Gulf of Mexico, and over the past decades has put so much fertilizer and pesticide laden water into the Mississippi estuary that there now exists what is called a "dead zone" of roughly 20,000 km^2 (almost the size of the state of New Jersey).

Bibliography

[1]   CropLife America. Website. (Accessed 6 December 2022, at www.croplifeamerica.org/).
[2]   Environmental Protection Agency, Pesticides. Website. (Accessed 6 December 2022, at epa.gov).
[3]   Crop Life Europe. Website. (Accessed 6 December 2022, at croplifeeurope.eu).
[4]   Fertilizer Australia. Website. (Accessed 6 December 2022 at https://fertilizer.org.au).
[5]   The Fertilizer Institute. Website. (Accessed 6 December 2022, at https://tfi.org).
[6]   Fertilizers Europe. Website. (Accessed 6 December 2022, at https://www.fertilizerseurope.com/).
[7]   International Fertilizer Industry Association. Website. (Accessed 6 December 2022, at https://www. fertilizer.org/).
[8]   Pestnetwork.com. Website. (Accessed 6 December 2022, at https://pestnetwork.com).
[9]   USGS Mineral Commodity Summaries 2022, Downloadable at: https://pubs.er.usgs.gov/publication/ mcs2022.

18 The paper industry

18.1 Introduction: the chemical composition of wood

The paper industry certainly is a major consumer of some basic commodity chemicals, but of course, wood remains the central commodity used in paper production. Wood is composed of cellulose, hemicellulose and lignin. When wood is used for the production of paper, these materials must be broken down chemically and reconfigured, so that fibers can be reconfigured and made into a single, uniform material without defects or weak points. Hardwoods and softwoods differ in the amounts of cellulose, hemicellulose and lignin that each contains, but this is usually not the driving factor in determining what wood will be used to produce paper. While virtually all wood can be used for producing pulp and paper, as can some other woody materials such as bamboo, the determining factor in choosing is often simply the growth rate of the trees as well as the price of the wood, and thus a steady supply of wood for the pulping process. For example, pine and birch grow quickly, and routinely can find use in paper production. Oak and maple are slower growing trees whose wood is more valuable in the production of furniture, housing interior paneling, and veneer, and thus are not normally used in paper production.

18.1.1 Cellulose

Cellulose is a linear organic polymer present in the cell wall of plants, with molecular weights as high as 100,000, and present in large amounts in wood, roughly 50 %. Figure 18.1 shows the repeat unit of cellulose. When wood is prepared for pulping by chipping, the bark is removed first, since it contains little cellulose that can be used. Rather, the bark is used to generate heat which is needed as the cellulose is broken down chemically.

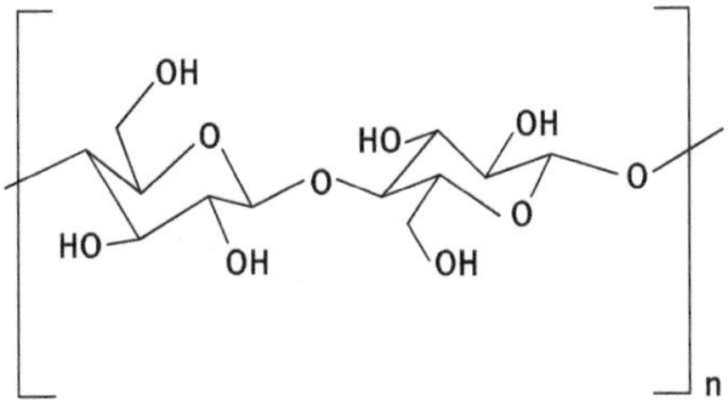

Figure 18.1: Cellulose.

https://doi.org/10.1515/9783110671094-018

18.1.2 Hemicellulose

Hemicellulose is a polymer akin to cellulose, but with a more random pattern, as well as a smaller general molecular weight. It is present along with cellulose in plants, and thus is incorporated into paper as the starting material is chipped and pulped.

Both cellulose and hemicellulose have thousands of sugar monomers linked covalently, and sharing a network of hydrogen bonds from one polymer chain to another. Both the intermolecular forces, as well as some of the covalent linkages in each, must be broken during the formation of pulp from wood chips.

18.1.3 Lignin

Lignin is also found in cell walls, and is an organic polymer with molecular weight of roughly 10,000 (smaller than both cellulose and hemicellulose), although this is difficult to state with certainty, because of the almost random nature of the repeat units within it. The material is highly aromatic, and the most difficult of the three components of wood to break down. Up to 30 % of the mass of wood can be lignin, so it is important that the chemical processes that break down wood be able to react with this material.

18.2 Virgin papers

The only requirement for a paper to be called 'virgin' is that it must be made directly from wood or annual plants without having any re-used or recycled paper in it. There are several trade organizations in North America and Europe devoted to the manufacturing of paper and wood products [1, 5, 6, 3], and they have developed an informal trade nomenclature to ensure paper types are described in an unambiguous manner.

18.3 Kraft process (a. k. a. the sulfate process)

This process, named from the German word "Kraft," meaning strong, and unrelated to the name of the food production company, breaks down wood to wood pulp that consists almost entirely of cellulose. The process separates the cellulose from the lignin and the pulp can then ultimately be used to produce paper. The steps involved in the process are as follows:

1. Impregnating the wood fibers. The wood chips are fed into a digester to a final ratio that is roughly 20 % wood and 80 % water.
2. Pressure digestion. This involves cooking the wood chips in what is called white liquor, a mixture of $NaOH$, Na_2S, and $NaSH$. The temperature is approximately 175 °C at 7–9 atm, for a 2–5 hour contact time.

3. Liquor recovery. This is the separation of the pulp which will be used for paper, paperboard or cardboard from the remaining liquid, now called black liquor (or sometimes brown liquor).
4. Blowing and volatiles collection. The liquor is concentrated and volatiles captured, so that the materials within the liquor can be re-oxidized and ultimately re-used.
5. Screening. This is a sieving process to remove material that is still too large to be properly part of the pulp used for paper.
6. Cellulose washing.
7. Bleaching. These two steps ensure the uniformity and purity of the remaining cellulose, and ensure no lignin remains in the pulp.

Overall, this process extracts most of the lignin from the wood, leaving good pulp with strong fibers that make high grade paper.

This list of steps does not properly reflect the amount of chemicals required in the process. The Kraft process is almost entirely a closed system, although sodium sulfate does have to be added at times. The reaction chemistry is not stoichiometric, and thus is difficult to represent with simple equations. Table 18.1 lists the chemicals required for Kraft processing of wood pulp.

Table 18.1: Chemicals Required in the Kraft Process.

Chemical	Processing step
Water	Steam heating of the wood chips
White liquor	Digestion, break down of the lignin
Black liquor	Non-pulp portion of the mixture, containing organics. Is recovered, separated, and re-used
Sodium sulfate	Production of sodium sulfide, Na_2S, for S^{2-} degradation of lignin
Calcium carbonate	Production of CaO, used to produce $Ca(OH)_2$, used to generate Na_2S

The steam produced in the process is also used to generate electricity, as is bark that has been removed prior to the chipping of wood, making many Kraft Process mills self-sufficient in electrical energy use.

18.3.1 By-product, lignin

Lignin extracted from the wood during the Kraft Process is also used. Routinely, it is burned on site as a source of energy – often by producing steam which is run through turbines – that is used to operate other parts of the process.

In the last fifteen years, lignin has also been used as a substitute for various injection-moldable plastics. The chemistry involved in "plasticizing" the lignin remains proprietary, but when the object from which it is made is no longer needed or usable, it can be discarded, since the material is essentially a bio-degradable wood product [7].

18.3.2 Tall oil

What is called tall oil is extracted from the black liquor. It is a mixture of organic materials, usually of much lower molecular weight than cellulose or lignin, and finds use as an adhesive, an emulsifier, or a component in inks. Because different trees produce black liquors with different compositions, the amounts of tall oil per batch of pulp differ. The general range is 25–50 kg per ton of pulp.

Through this all, it can be seen that the production of paper requires a large input of chemicals, water, and energy.

18.4 High quality papers – acid sulfite process

The sulfite process involves the production of pulp that is higher in cellulose, and that has removed more of the lignin from the final product. Whereas the Kraft Process utilizes sulfates, the Sulfite Process utilizes metal sulfites or bisulfites, ultimately made from sulfur. The reaction chemistry can be simplified to the following:

$$S + O_2 \longrightarrow SO_2$$
$$SO_2 + H_2O \longleftarrow \longrightarrow H_2SO_3$$

Burning the sulfur, then absorbing it in water, must be controlled carefully, to avoid over-oxidation to sulfur trioxide, which would lead to aqueous solutions containing sulfuric acid. In this process, sulfurous acid is the product required to produce the sulfite and bisulfite salts. The reaction chemistry for this can be represented as follows:

$$NaOH + H_2SO_3 \longrightarrow NaHSO_3 + H_2O$$

And then:

$$NaHSO_3 + NaOH \longrightarrow Na_2SO_3 + H_2O$$

Sodium is not the only cation that can be used in the process. Ammonium and potassium are also suitable.

As with the Kraft Process, in the Sulfite Process wood chips must be in contact with this white liquor in a digester for several hours (5–15 hours). This provides the time to effect lignin breakdown.

The separation of the pulp from what is called brown liquor or red liquor is a matter of washing the pulp and concentrating the contents of the liquor for re-use. Since this process does not break lignin down as severely as the Kraft Process, the pulp by-products can be used in other applications.

18.5 Steps in paper manufacturing

Virtually all developed countries have several companies that own and operate paper mills. The paper making industry is ubiquitous enough that there are mills and companies devoted exclusively to niche paper products, and high end products that are used widely, but in smaller commodities, such as waxed papers, or commodity package labeling [7].

While there are different processes for producing paper based on the end use for the material, there are also some basic steps that essentially all papers go through when made on an industrial scale. Starting with wood pulp that has been processed these steps include the following:

1. Crushing and refining. This step is designed to produce fibers of relatively similar and small lengths, so that the final product has the strength and feel that is required of it.
2. Coagulation and fiber coating. Aluminum sulfate, often called papermaker's alum, is mixed with the material, generally as a slurry.
3. Fillers. Fillers are added to give paper a specific look and feel. In general, fillers should not degrade the strength and durability of the paper. Titanium dioxide as well as calcium carbonate are two fillers that are also bulk commodity chemicals we have discussed in earlier chapters. There are other materials that can be used in this regard as well.
4. Sizing. This is the incorporation of materials into the paper that slows absorption of water into the finished product. Traditionally, a variety of soaps or waxes have been used, but today, rosins or various succinic anhydrides are used. This imparts the proper surface characteristics that the finished paper will hold ink on it long enough for the ink to dry without being soaked into the paper.
5. Wet strength resins. The addition of this class of materials increases the strength of paper when they are wet. Urea-formaldehyde resins have been used for decades, but polyamide-epichlorohydrin resins are also used [2].
6. Dyeing. Perhaps obviously, dyes are added to produce papers of specific color [5]. The types of dyes are numerous, and have been discussed earlier in the book. Adding a dye to a paper batch is much like adding a filler, in that the dye can change the mechanical properties of the paper. It should not degrade the strength of the resultant material.

18.6 Paper uses

The uses for paper and paperboard are numerous. Besides books, newsprint, writing paper, bags and boxes, there are such applications as paper plates and food packaging, and wallpaper or shelf paper, as well as labeling for consumer end use bottles, cans, and jars. Large cardboard shipping crates and containers are an industry that uses hundreds of thousands of tons of material annually.

Ultimately, this is an industry with amazing diversity in the end use products it manufactures. Niche manufacturing of special papers, inclusive of papers with vibrant colors, or special textures, makes this even more varied [4].

18.7 Paper re-cycling

Paper recycling has a long history, going back at least into the 1960s in many communities throughout the United States. Unfortunately, the geographic size of the United States has worked against the development of a single, national law or set of guidelines on the recycling of paper and paper-based materials, meaning paperboard and cardboard. States or local governments have been left to determine what level of paper recycling is considered suitable for their particular area. Thus, municipalities where waste disposal is costly often choose to institute a cardboard, paperboard, and paper recycling program, while communities that have plenty of room for landfills, and low costs to transport waste materials to them, do not.

18.8 Low grade papers

Recycled papers are usually of somewhat lower grade than the paper from which they came, and while in theory recycling of a material like this should be an indefinitely repeatable process, the production of paper continues to degrade the fibers, which means that there are usually no more than five or six cycles in the recycling of paper. Beyond this point, the paper quality is lower than that of newsprint, and thus it is not very durable. Such material can be used in various applications where it is not readily seen, such as housing insulation.

Bibliography

[1] American Forest and Paper Association. Website. (Accessed 6 December 2022, at www.afandpa.org/).
[2] Chernier, P. J. Survey of Industrial Chemistry, Second Edition, Wiley, 1992.
[3] Confederation of European Paper Industries. Website. (Accessed 6 December 2022, at www.cepi.org/).
[4] Dunn Paper. Website. (Accessed 6 December 2022, at www.dunnpaper.com/).

[5] Pulp and Paper Manufacturers Association of the USA. Website. (Accessed 6 December 2022, at http://www.naylornetwork.com/ppi-otw/articles/?aid=196402&issueID=27429).
[6] Pulp and Paper Technical Association of Canada, PAPTAC. Website. (Accessed 6 December 2022, at www.paptac.ca/).
[7] Tecnaro. Website. (Accessed 6 December 2022, at https://www.tecnaro.de/en/).

19 Pharmaceuticals

19.1 Introduction

This book has already treated the large-scale production of numerous chemical compounds, elements, and other materials and commodities, and made the point that almost all have substantially changed the world in some way. The large-scale production of medicines and pharmaceuticals, including everything from aspirin to heroin (still a legally prescribable narcotic pain killer in some countries), is another area in which the world has been drastically changed. Because of the availability of medicines in high purity and low cost, the average human life span in the developed world has increased by decades, certainly if one compares human lifetimes from the year 1800 for instance, until today. In addition, the advent of numerous medicines, drugs, and vaccines has decreased infant and child mortality enormously in many parts of the world.

Most books that discuss drug use and production treat the materials they examine by dividing the drugs into classes based on their chemical formulae, or based on their class of action when used. Here, we will simply look at pharmaceuticals in terms of production, and focus on a few well-known examples.

Prescription and over the counter (OTC) drugs are considered very low volume and high cost when we compare them to the basic commodity chemicals of the early chapters of this book. But drugs are potent materials (largely they are organic molecules that are stable in water, but not always) that in very small amounts affect a person with body mass far in excess of the dosage.

19.2 The top 100 prescription medicines sold

As mentioned, drugs and medications can be listed in a variety of ways. The website drugs.com chooses to compile a listing of prescription medications based on sales figures. Table 19.1 shows these sales for the final quarter of the year 2012 [3]. We have added a column that explains briefly for what each prescription medicine is generally used.

Putting the entire spectrum of prescription drug sales, and this rather large list into perspective against the major causes of death in the world is enlightening. The World Health Organization states on its web site:

"In high-income countries more than two thirds of all people live beyond the age of 70 and predominantly die of chronic diseases: cardiovascular disease, chronic obstructive lung disease, cancers, diabetes or dementia. Lung infection remains the only leading infectious cause of death.
In middle-income countries, nearly half of all people live to the age of 70 and chronic diseases are the major killers, just as they are in high-income countries. Unlike in high-income countries, however, tuberculosis, HIV/AIDS and road traffic accidents also are leading causes of death.

https://doi.org/10.1515/9783110671094-019

Table 19.1: Prescription Drug Sales and Uses.

Name	Manufacturer	Sales, $US (×1,000)	Used for:
Abilify	Otsuka Pharmaceutical	1,478,301	Diabetes
Nexium	Astra Zeneca Pharmaceutical	1,441,472	Acid reflux
Crestor	Astra Zeneca Pharmaceutical	1,275,483	High cholesterol
Cymbalta	Eli Lilly	1,227,484	Anxiety and depression
Humira	Abbott Laboratories	1,206,377	Arthritis and Crohn's disease
Advair Diskus	GlaxoSmithKline	1,204,874	Asthma and chronic obstructive pulmonary disease (COPD)
Enbrel	Amgen	1,088,533	Arthritis
Remicade	Centocor Ortho Biotech	935,584	Arthritis, Crohn's disease
Copaxone	Teva Pharmaceuticals	908,061	Multiple sclerosis
Neulasta	Amgen	803,286	Bone marrow loss
Rituxan	Genentech, Inc.	779,026	Lymphoma, leukemia
Spiriva	Boeringer Ingelheim Pharmaceuticals	702,246	COPD
Atripla	Gilead Sciences, Inc.	695,411	HIV infection
Oxycontin		668,415	Pain
Januvia	Merck & Co.	658,957	Diabetes
Avastin	Genentech	634,201	Cancers
Lantus	Sanofi-Aventis	599,359	Blood sugar level control
Lantus Solostar	Sanofi-Aventis	597,688	Blood sugar level control
Truvada	Gilead Sciences	565,853	HIV
Epogen	Amgen, Inc.	514,165	Anemia
Lyrica	Pfizer	511,751	Seizures, general anxiety disorder
Celebrex	Pfizer	505,850	Arthritis
Diovan	Novartis	474,603	High blood pressure
Herceptin	Genentech	464,737	Breast cancer
Namenda	Forest Pharmaceuticals	454,671	Alzheimer's disease
Gleevec	Novartis Corp.	436,298	Cancers
Suboxone	Reckitt Benkiser Pharmaceuticals	417,264	Pain, opioid addiction
Vyvanse	Shire US	408,526	Attention deficit hyperactivity disorder (ADHD)
Methylphenidate (Ritalin)		399,702	ADHD
Lucentis	Genentech	395,931	Vision loss
Enoxaparin		394,149	Deep vein thrombosis
Zetia	Merck & Co	375,281	Cholesterol
Avonex	Biogen Idec	371,884	Multiple sclerosis
One Touch Ultra		349,860	Blood glucose monitor
Lidoderm	Endo Pharmaceuticals	348,610	Skin inflammation
AndroGel	Abbott Labs	346,491	Low testosterone
Symbicort	AstraZeneca Pharmaceuticals	340,401	Asthma, COPD
Rebif		311,851	Multiple sclerosis
Novolog	Novo Nordisk	311,150	Insulin levels
Levemir	Novo Nordisk	308,726	Insulin levels

Table 19.1 (continued)

Name	Manufacturer	Sales, $US (×1,000)	Used for:
Alimta	Eli Lilly	298,106	Cancers
Novolog FlexPen	Novo Nordisk	293,569	Insulin levels
Viagra	Pfizer	290,880	Erectile dysfunction
Seroquel XR	AstraZeneca Pharmaceuticals	284,989	Schizophrenia
ProAir HFA	Teva Pharmaceuticals	284,649	Asthma, COPD
Synagis	MedImmune, Inc.	281,760	Respiratory infections
Flovent HFA	GlaxoSmithKline	268,626	Anti-inflammatory
Niaspan	Abbott Laboratories	268,151	Cholesterol
Amphetamine/dex-troamphetamine		267,476	ADHD, narcolepsy
Incivek	Vertex Pharmaceuticals	266,240	Hepatitis C
Lovaza	GlaxoSmithKline	266,067	High triglyceride levels
Budesonide		265,773	Asthma, allergies
Combivent	Boehringer Ingelheim Pharmaceuticals	265,042	Asthma, COPD
Modafinil		264,781	Alertness
Nasonex	Merck & Co.	260,739	Inflammation
Humalog	Eli Lilly	257,939	Low insulin
Acetaminophen/hydrocodone		248,050	Pain
Procrit	Janssen Pharmaceuticals	246,708	Anemia
Metoprolol		239,668	Hypertension
Isentress	Merck & Co.	238,849	HIV infection
Valsartan		237,796	High blood pressure
Reyataz	Bristol-Myers Squibb	237,659	HIV infection
Cialis	Eli Lilly	237,294	Erectile dysfunction
Tamiflu	Roche Pharmaceuticals	237,097	Anti-viral
Restasis	Allergan, Inc.	233,071	Immunosuppressant
Gilenya	Novartis Corp.	232,746	Multiple sclerosis
Neupogen	Amgen	231,017	Bone marrow stimulation
Victoza	Novo Nordisk	227,520	Type-2 diabetes
Janumet	Merck & Co.	222,968	Diabetes
Zostavax		222,633	Shingles
Vytorin	Merck & Co.	221,891	High cholesterol
Prezista	Janssen Pharmaceuticals	216,609	HIV infection
Aciphex	Eisai Corp.	213,378	Ulcers
Orencia	Bristol-Myers Squibb	211,272	Arthritis
Dexilant	Takeda Pharmaceuticals North America	210,175	Gastro-oesophageal reflux disease
Aranesp	Amgen	210,027	Anemia
VESIcare	Astellas Pharma	209,122	Overactive bladder
Fenofibrate		206,565	High cholesterol
Pradaxa	Boehringer Ingelheim Pharmaceuticals	205,238	Stroke prevention

Table 19.1 (continued)

Name	Manufacturer	Sales, $US (×1,000)	Used for:
Prevnar 13	Wyeth	204,585	Pneumococcal vaccine
Adderall XR	Shire US	202,948	Attention deficit hyperactivity disorder
Betaseron	Bayer Healthcare Pharmaceuticals	202,032	Multiple sclerosis
Renvela	Genzyme Corporation	201,365	Hyperphosphatemia
Atorvastatin		196,153	High cholesterol
Fentanyl		194,261	Pain
Benicar	Daiichi Sankyo	193,275	High blood pressure
Lunesta	Sunovion Pharmaceuticals	192,034	Insomnia
Synthroid	Abbott Laboratories	190,754	Hypothyroidism
Evista	Eli Lilly	189,910	Osteoporosis
Ventolin HFA	GlaxoSmithKline	186,276	Asthma, COPD
Xolair		182,829	Allergic asthma
Sensipar	Amgen	182,611	Hyperparathyroidism
Stelara		182,409	Arthritis
TriCor	Abbott Laboratories	180,903	High cholesterol
Xeloda	Roche Pharmaceuticals	178,469	Breast and colorectal cancer
Xgeva	Amgen	173,788	Osteoporosis
Zyvox	Pfizer	169,211	Infections
Focalin XR	Novartis Corp.	168,218	ADHD
Pioglitazone		165,901	Diabetes
Sandostatin LAR Depot	Novartis Corp.	162,139	Inhibit growth hormone

In low-income countries less than one in five of all people reach the age of 70, and more than a third of all deaths are among children under 15. People predominantly die of infectious diseases: lung infections, diarrhoeal diseases, HIV/AIDS, tuberculosis, and malaria. Complications of pregnancy and childbirth together continue to be leading causes of death, claiming the lives of both infants and mothers." [10]

It is noteworthy that the top 100 list does not include any drugs that treat malaria or tuberculosis, or any of the other diseases that most afflict people in the developing world. Arguments have been made by advocacy groups in the past that drug companies manufacture medicines that can be marketed in the developed world where people can pay for them, while people continue to die of much more common diseases in what is called the Third World, where medications are either not sold or are too expensive for the population. Malaria and anti-malarial drugs can serve as a good example of this, as anti-malarial drugs are given to United States armed forces personnel when they deploy to parts of the world where the disease is still common. Thus, anti-malarial drugs certainly exist, but do not seem to be sold extensively where malaria is a killer.

The argument is then that the manufacturers of pharmaceuticals are simply targeting a population that can pay for their medications. While this appears damning, it should be remembered that pharmaceutical manufacturers, like other companies, exist to make a profit through the sales of their end products. Logically, they must sell to customers who have the money to pay, which generally means people in the developed world.

19.3 Over the counter drugs

Over the counter drugs, often abbreviated OTC, have been proven to be non-habit forming, and safe to the public when the directions for use are followed. Some have long histories of use, such as aspirin. Others have only been on drug stores shelves for a few years, or decades at most, such as the decongestant pseudoephedrine.

The Consumer Healthcare Products Association categorizes OTC sales [2], but does not break them down by product and manufacturer. A list of the highest selling OTC items includes those items in Table 19.2.

Table 19.2: OTC Item Sales.

OTC Category	2012 Sales $US (×1 M)
Acne remedies	624
External analgesics	516
Internal analgesics	3,893
Antidiarrheals	216
Anti-smoking	1,153
Cough and cold medicines	6,631
Eye care	848
First aid	1,106
Foot care	589
Heartburn and anti-gas	2,268
Laxatives	1,358
Lip remedy	819
Oral antiseptics	1,394
Sun blocks and sunscreens	1,005
Toothpaste	2,449

While it may seem odd to have mouthwashes and toothpastes included on such a list, these products are now marketed as means to maintain good dental hygiene, and thus are preventatives when it comes to diseases of the mouth.

Clearly, the manufacture and sale of prescription medications and over-the-counter medications is a huge industry. The dollar values in each table, when added up, result in a bottom line that is a multi-billion dollar industry in both cases.

19.4 Drug synthesis

The synthesis of almost all drugs involves several steps in what is now considered classical organic chemistry, and may involve steps that are proprietary to a specific company, or a process that is patented.

19.4.1 Synthesis of common medicines

Aspirin

Aspirin has been a common pain reliever for over 100 years, and is still sold in large quantities today, despite competition from other OTC pain relievers. It is now synthesized from phenol, as shown in Figure 19.1, although it was originally found in willow tree bark (sufferers used to chew the bark to relieve pain). Thus, since phenol is the main starting reactant, aspirin today ultimately comes from oil.

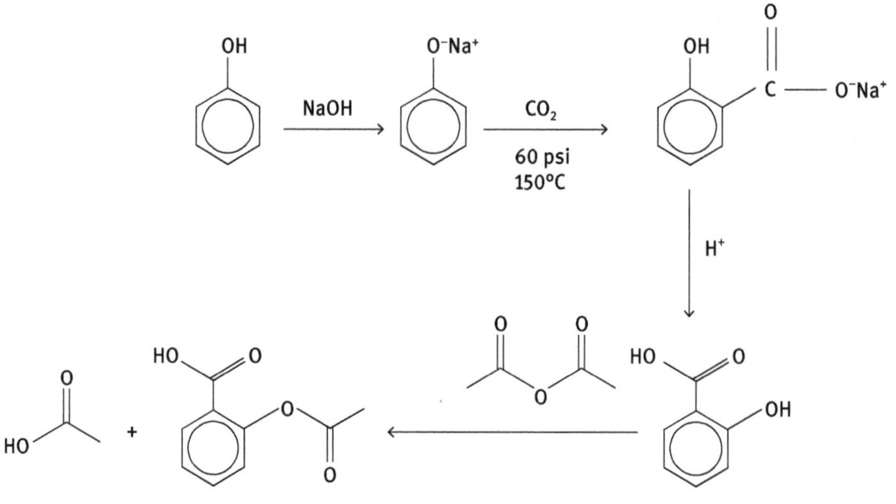

Figure 19.1: Aspirin Synthesis.

Acetaminophen

Like aspirin, acetaminophen is an OTC medication taken to relieve pain. Also like aspirin, acetaminophen synthesis, as shown in Figure 19.2, begins with phenol.

Ibuprofen

Ibuprofen is another pain reliever that is sold over the counter, and that competes with both aspirin and acetaminophen in terms of sales. A synthesis of ibuprofen is shown in Figure 19.3. It can be noted that like the other two, ibuprofen now uses starting materials that are essentially derived from petroleum.

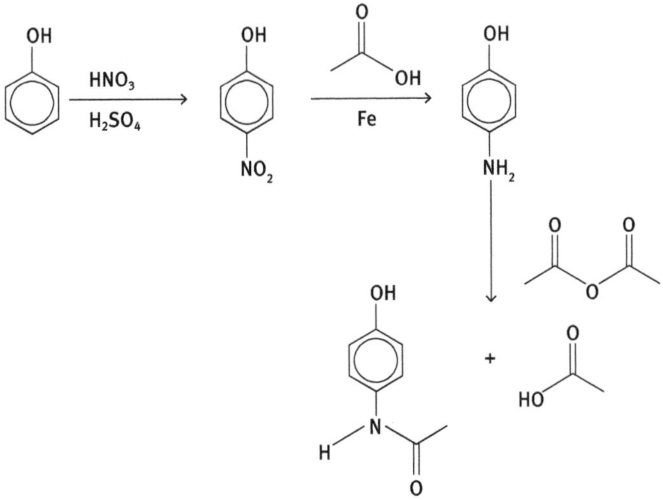

Figure 19.2: Acetaminophen Synthesis.

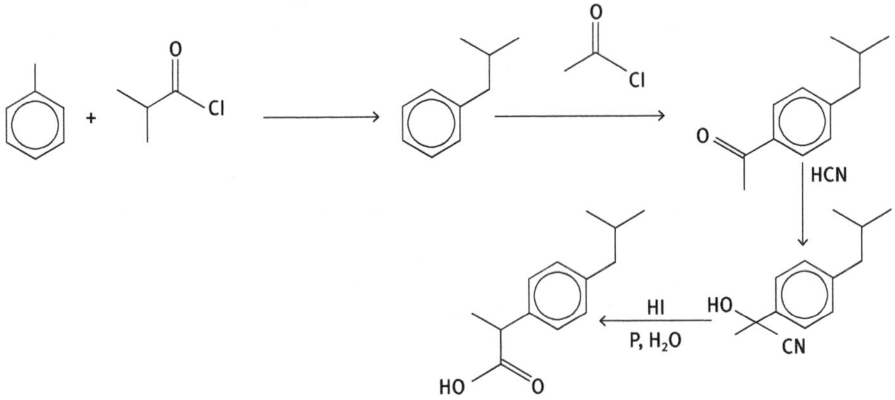

Figure 19.3: Ibuprofen Synthesis.

Codeine

Codeine is closely related to morphine, and the one step synthesis from morphine to codeine is shown in Figure 19.4. Codeine is actually found in opium poppy exudate, but in low enough quantities that it is easier to synthesize it from morphine (which is present in larger amounts). The major uses for this pharmaceutical is pain relief and cough suppression, although there are several other uses as well.

Morphine

Morphine is extracted from opium poppies. It is also used in the synthesis of codeine, above, where a single methyl group needs to be added to produce the codeine end prod-

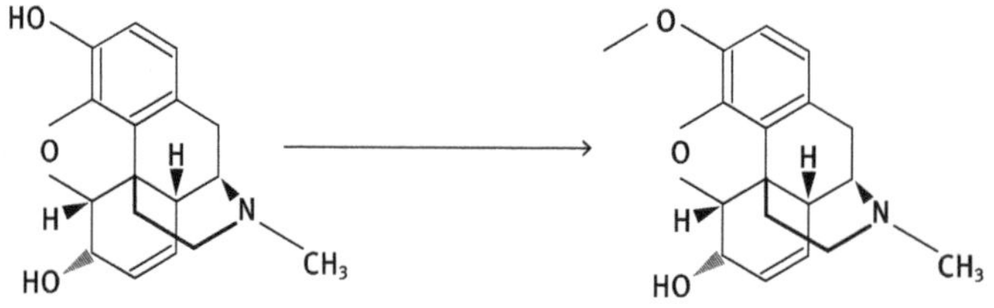

Figure 19.4: Codeine Synthesis.

uct. Morphine remains a tightly controlled substance, while as mentioned, codeine is a widely used medication today.

19.5 Recycling

The recycling of drugs and medications remains in its infancy in many parts of the developed world. Even though a number of national agencies and organizations exist that monitor and provide information about pharmaceuticals use and interactions with the greater environment [8, 6, 5, 1, 4], they do not necessarily provide instructions for the proper disposal and neutralization of old or outdated medicines. Since drugs and medicines are a class of consumer end use product, ultimately the information about proper disposal and the end of the life cycle need to be provided to the users.

The last decade has seen a number of studies about pharmaceutical materials that have ended up in waterways, usually because they were flushed down toilets, and the resulting water purification plants were unable to process them out of the water. Indeed, water purification plants are designed to treat sewage and soluble or insoluble inorganics, and have never been designed to remove water-soluble organics, meaning drugs, from the water stream. The result in some cases has been lakes or rivers in which aquatic life has been feminized, since a number of potent pharmaceuticals are estrogen mimics [9]. On a promising note, in at least one case a lake has been returned to its pre-pharmaceutically polluted state.

Pharmaceutical recycling programs that do exist are not necessarily run by municipalities. Rather, drug store chains have specific times and store locations where they take back unused or outdated medications. Stores and companies that do have such programs generally advertise this on their corporate web site. As well, there are now some services that accept medications for proper destruction and disposal [7].

Bibliography

[1]   American Pharmaceutical Manufacturers Association. Website. (Accessed 6 December 2022, at https://phrma.org/).
[2]   Consumer Healthcare Products Association. Website. (Accessed 6 December 2022, at www.chpa.org/).

[3]   Drugs.com. Website. (Accessed 6 December 2022, at www.drugs.com/).

[4]   European Federation of Pharmaceutical Industries and Associations. Website. (Accessed December 2022, at www.efpia.eu/).

[5]   European Medicines Agency. Website. (Accessed 6 December 2022, at https://www.ema.europa.eu/en).

[6]   Health Canada. Website. (Accessed 6 December 2022, at www.hc-sc.gc.ca/index-eng.php).

[7]   Pharmaceuticals Return Service. Website. (Accessed 6 December 2022, at www.pharmreturns.net/).

[8]   United States Food and Drug Administration. Website. (Accessed 6 December 2022, at www.fda.gov/).

[9]   Water Reuse: Potential for Expanding the Nation's Water Supply Through Reuse of Municipal Wastewater. National Academies Press, downloadable at www.nap.edu.

[10]   World Health Organization. Website. (Accessed 6 December 2022, at https://www.who.int/health-topics/medicines#tab=tab_1).

20 Surfactants and detergents

20.1 Introduction and historical production

Historically, the production of soaps and detergents has been as a part of animal husbandry and the use of animal parts that would not normally be used for human consumption, such as fats. The first soaps were a mixture of fats and wood ashes. The production of such soap was often done in the home, and if such soaps were not used quickly enough, the material began to decompose.

The terms soap, surfactant and detergent are used somewhat interchangeably, although there are some differences. Detergents are considered more soluble in hard water than soaps, because they tend to contain either an ammonium salt or a sodium or potassium cation. What all have in common is one or more of a wide variety of hydrophobic tail groups, and some form of polar head group. All surfactants are not necessarily ionic, but all function because of their molecular bi-polarity. Figure 20.1 shows a simplified schematic, using only straight chain hydrophobic tails (hydrophobic portions can also be branched). From top to bottom, the figure shows anionic, cationic, zwitterionic, and non-ionic surfactant types.

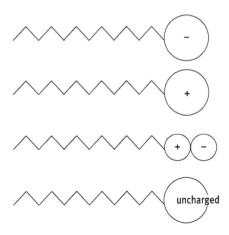

Figure 20.1: Surfactant Types.

Despite the massive nature of the surfactants and detergents industry – it produces billions of pounds of materials per year, and is a multi-billion dollar industry – artisanal soap-making has maintained a niche market, and even taken an enhanced market share in the past few years. Plenty of small businesses produce soaps and other beauty and cleaning products, with the specific intent of marketing themselves to the high end, luxury market, in which consumers will pay somewhat more for products that are organic in manufacture, or made with all natural materials and components [4].

https://doi.org/10.1515/9783110671094-020

The United States Census has a portion devoted to such products, under the broad heading of manufacturing. These included until 1997 what were called standard industrial code descriptions (SIC) 2840 for soap, cleaners, and toilet goods, as well as the related SIC2841 Soap and other detergents, SIC2842 for polishes and sanitation goods, and SIC2843 for Surface active agents [8, 1]. The SICs were subsequently replaced by the North American Industry Classification System (NAICS), shown in Table 20.1 by number [8].

Table 20.1: NAIC System for Surfactants.

Category	NAICS Code
Soap and Cleaning Compound Manufacturing	32561
Soap and other Detergent Manufacturing	325611
Polish and other Sanitation Good Manufacturing	325612
Surface Active Agent Manufacturing	325613
Soap, Cleaning Compound, and Toilet Preparation Manufacturing	3256

We have not used the NAICS for any other subject we have included in the book, even drugs and pharmaceuticals covered in Chapter 19, simply because none of the other materials and processes lend themselves as well to user end products and manufactured goods that are tracked by the US Census. The Census tracks business types, and these materials are often directly associated with specific businesses. Here, the NAICS system breaks the products into subcategories, which is the purpose of the six digit code. The four digit code 3256 encompasses all the soaps and cleaning compounds. Using this system, the US Census is able to determine every ten years what portion of the nation's economy is dependent upon the use of surfactants, soaps, and detergents.

But the US Census is hardly the only organization tracking the use, and in some cases promoting the uses, of surfactants and detergents [1, 5, 3]. All non-governmental organizations that are devoted to this industry take pains to explain not only the uses of such products, but the responsible uses. In the past, environmental degradation has been caused by excessive and improper uses of surfactants and detergents. In the last thirty years, the population of the world has increased significantly, which means that responsible use of personal and industrial cleaning products should be undertaken, since environmental damages in one area often translate to drinking water problems for downstream communities. The American Cleaning Institute issues an annual Sustainability Report that reiterates proper use of such materials [1].

20.2 Current syntheses

As mentioned, surfactants can be divided into four broad categories, based on their molecular structures. These are anionic and cationic, zwitterionic, and non-ionic surfactants. Within each of these categories, the length and branching of the hydrophobic chain tail can be linear or branched, and saturated or unsaturated. For the purposes of explanatory figures, we will generally represent molecules with linear, hydrophobic tails.

The starting materials for most surfactant production are triglycerides, rendered from either animal tallow or plant sources, depending on the desired end use product. Animal tallow is usually obtained from meat packing factories, while plant matter has a variety of sources.

The reaction to break down the triglycerides is carried out at elevated temperature and pressure, possibly with a metal oxide or acid catalyst, and produces fatty acids and glycerol, as shown in Figure 20.2.

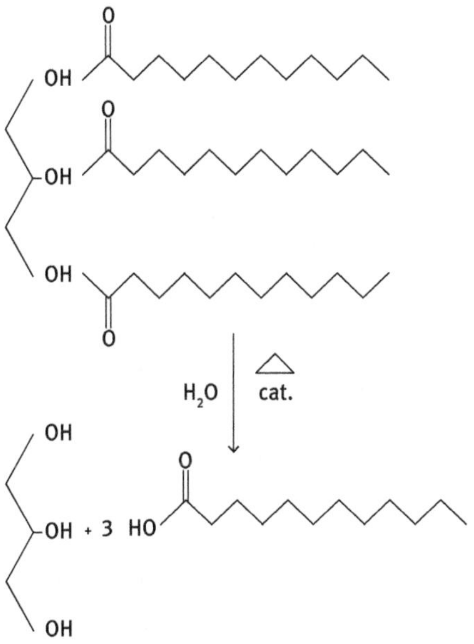

Figure 20.2: Triglyceride Breakdown.

Although we are focusing on the surfactant molecules, glycerol is also recovered and used in a variety of products.

20.2.1 Anionic surfactants

Anionic surfactants compose the largest sector of the industry, and usually have either a carboxylate or a sulfate polar head group. The counter cations for such materials are often alkali metals that were introduced during a saponification. They are produced by the saponification of fatty acids with a hydroxide (the industry tends to refer to sodium or potassium hydroxide by the industrial name 'lye'). The end result is a fatty acid salt and water, as shown in Figure 20.3, using sodium hydroxide.

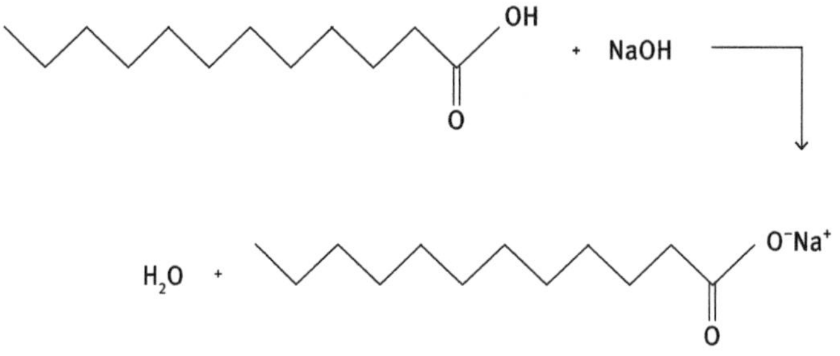

Figure 20.3: Surfactant Production.

20.2.2 Cationic surfactants

As mentioned above, the major difference between cationic and anionic surfactants is the head group's polarity. Cationic detergents have quaternary ammonium head groups almost exclusively. The nitrogen is thus positively charged, as shown in Figure 20.4.

Figure 20.4: Cationic Surfactant.

The reason fewer cationic surfactants are used in household applications is simply that many greases, oils, and other dirt that people come into contact with have been found to react better with anionic surfactants. But cationic surfactants still find use in non-household applications, specifically in use as fuel additives. Using amounts in the 100 ppm range, amine-based detergents in fuels help prevent fouling of the metal components.

20.2.3 Zwitterionic or di-ionic surfactants

As the name implies, zwitterionic surfactants possess both a negative and positive charge, in a 1:1 ratio. They are not as large a market share of soaps, surfactants, and detergents as are anionics, but they can interact either through the positive or the negative end depending upon the pH of the solution in which they are solvated.

20.2.4 Non-ionic surfactants

Again, as the name implies, non-ionic surfactants do not contain ionic pairs in the actual make-up of the detergent material. They possess a non-polar tail group, and a head group that is polar, but not ionic. Several of the very long chain alcohols function well as non-ionic surfactants. Examples are shown in Figure 20.5. Both stearyl alcohol and cetyl alcohol exist as white, waxy solids at room temperature. The term alcohol is used to describe small organic liquids so often that some explanation here is in order. At these higher molecular weights, the alkyl portion of the molecule dictates the physical state and properties of it. This is convenient within the surfactant industry, because stearyl alcohol, as well as the others, can be used not only as surfactants, but as emulsifiers and thickeners. This is important in producing end use products with the look and feel that customers want.

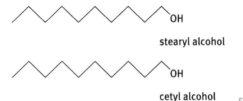

stearyl alcohol

cetyl alcohol   Figure 20.5: Non-Ionic Surfactants.

20.3 Function of surfactants

The function of surfactants is not to "clean clothes," as is often mis-perceived. Rather, the function of a surfactant is to encapsulate the dirt in clothing, or foreign material on an object, in a micelle which has the hydrophobic ends surrounding the dirt, and the hydrophilic heads pointed out towards the water, as shown in Figure 20.6. In the case of cleaning clothing, once these micelles have formed, the agitator in the washing machine mechanically beats the clothing enough that the dirt-surfactant-water solution is carried away from the clothes. In the case of dish washing, the sprayer-agitator functions in the same way.

20.4 Additives

Numerous additives find their way into soaps, surfactants and detergents. These include the following, but do not represent a complete list:
- Colors
- Fragrances
- Bleach
- Water softeners

Figure 20.6: Micelle Formation.

- Grits
- Boosters
- Fillers
- Probiotics
- Trisodium phosphate (TSP)

The reasons for such additives are varied. Colors are added often simply to make the end use product attractive to the consumer. Bright colors tend to stand out more on store shelves.

Fragrances are added to enhance the appeal of the product, and help the buyer believe the material is more natural (although there is a very old joke that asks if lemon scented soaps clean better, why do we all not simply clean ourselves, our clothes, and our dishes with lemons?).

Bleach is often added because while the soap removes dirt by forming micelles and being washed away, white clothing is not made whiter by that process. Bleach whitens the fibers.

Water softeners are added to ensure that scale and other non-soluble material does not build up on the inside of water lines and pipes. The softener helps precipitate the ions out of solution, so they can be carried into the waste stream.

Grits are added to give dish washing formulas what is called scrub-ability, to help loosen large particles of food and other material from dishes and silverware.

Boosters are advertised as being able to increase the detergent's ability to clean, which translates into a greater ability to complex dirt, grease, and oils in fabrics. The famous 20 Mule Team Borax is still marketed as a laundry detergent booster.

Fillers such as sodium sulfate are added to bulk up the end material to a point where the portion size "feels right" to the consumer. Basically, if a person feels that 150 mL of cleaner (about half a cup) is the right portion size, but all that is needed for a washer load of clothing is 50 mL, they will still add 150 mL, even though that is three times the amount of product required. The addition of filler dilutes the product without degrading its action, and prevents the release of excess soap or detergent into the water stream.

Probiotics are added to some shampoos and soap with the thought and belief that they can produce a product that is healthier and fights environmental bacteria. Unfortunately, the prefix "pro" appears to be used to mislead members of the public into believing the product is in some way professional or of professional strength. This is never the case, for shampoos or for foods. The prefix simply means beneficial, for example: pro-biotic, versus anti-biotic. The European Food Safety Authority has found claims for probiotics related to skin care products to be unsupportable [2], and the US Food and Drug Administration continues to study whether or not there is merit in them in these claims [7].

Trisodium phosphate, or TSP, is mentioned here because of its wide use historically. It has essentially been phased out because phosphates that are flushed into the waste water stream, and ultimately into freshwater bodies in the environment have been responsible for large algal blooms, the subsequent oxygen loss in the water, and significant environmental damage. TSP is still manufactured from phosphate rock on an industrial scale, and finds limited use in cleaning surfaces painted with oil-based paints, but it finds use now as a flux in welding copper tubing.

20.5 Artisanal soap making

As mentioned in the introduction, above, soaps can still be made at home, in small batches, as a modern day type of cottage industry [4, 6]. Most artisanal soap makers use plant matter as the source of the fatty acid, and not animal tallow. Examples include olive oil, coconut oil, palm oil, and castor oil. The amount of oleic aids varies in each, and from one batch to another, but all can undergo a saponification reaction.

20.6 Recycling

Unfortunately, while the surfactant and detergent industry produces all formulations with the knowledge that these materials will often be used either residentially or in commercial or industrial settings, and thus will be flushed from homes and businesses into local waste water streams, there is little that can be done to make such materials easy to remove from water in a waste water treatment plant. Soaps, surfactants, and detergents are produced precisely so that they are miscible with water, and so that they will stay dissolved in it.

Because these are completely end use products, there are no recycling programs for such materials. Waste water treatment plants, discussed in Chapter 8, do clean water of such materials – which is referred to in such context as pollutants – but are designed to focus on the water, not recovery of the materials from the water.

Bibliography

[1] American Cleaning Institute. Website. (Accessed 7 December 2022, at www.cleaninginstitute.org/).

[2] European Food Safety Authority. Website. (Accessed 7 December 2022, at www.efsa.europa.eu/).

[3] International Network of Cleaning Product Associations. Website. (Accessed 7 December 2022, at www.incpa.net/index.html).

[4] Lush Life. Website. (Accessed 7 December 2022, at www.lushusa.com/).

[5] Soap and Detergent Association of Canada. Website. (Accessed 7 December 2022, at https://opengovca.com/corporation/598313).

[6] How Soap Is Made Website. (Accessed 7 December 2022 at https://www.youtube.com/watch?v=Kc7duzDEa6Y).

[7] US Food and Drug Administration. Website. (Accessed 7 December 2022, at www.fda.gov).

[8] US Government Census. Website. (Accessed 7 December 2022, at https://www.census.gov/search-results.html?q=detergents&page=1&stateGeo=none&searchtype=web&cssp=SERP&_charset_=UTF-8).

21 Rubber

21.1 Introduction

Few organic materials that are used on a large scale have defied chemical synthesis for as long as rubber. While people have known of rubber for centuries, and have used it in a wide variety of applications, natural rubber has long been a material of limited use, in that it becomes sticky when heated, and brittle when cold. As well, for centuries, the removal of trees, seedlings, or seeds of the rubber tree from the Amazon basin was illegal under the laws of Brazil [8, 9]. A rather amazing, accidental discovery, now called vulcanization, allowed the improvement of the working qualities of rubber [9], and the smuggling of seeds from the Amazon and export of them throughout the world changed the world's economy when it comes to this substance [8].

21.2 Sources

21.2.1 Natural rubber

The major worldwide source of rubber remains the Hevea brasiliensis plant, commonly called the rubber tree. As the Latin name implies, the plant grows naturally and wild in Brazil, and the restriction on its export was jealously guarded for decades in the nineteenth century. Seeds for the tree were eventually smuggled out of Brazil in the year 1876 by the Englishman Henry Wickham, who brought them to Kew Garden [8]. Because of this, more than 90 % of today's natural rubber is grown in southeast Asia, where the soil and rainfall are best for the plant's cultivation, where rubber plantations have been successfully cultivated, and where Britain was the colonial power when the trees were introduced [7]. Attempts to cultivate large rubber plantations in the Amazon basin have generally been less successful because of the aggressive growth of the surrounding jungle.

One of the larger attempts to cultivate rubber trees in the Amazon by a foreign company was that of the Ford Motor Company in 1928. An area roughly the size of Connecticut (10,000 km^2) was purchased and named "Fordlandia." Areas of jungle were cleared, people and equipment were moved in, and cultivation was begun. Unfortunately, it proved unsustainable, and the land was sold back to the Brazilian government in 1945, the same year that synthetic rubber became widely available [3].

Harvesting raw rubber involves making scores on the tree's trunk with a large knife, then collecting the exuding sap, often called latex. This process does not kill the tree, and thus can be repeated numerous times, simply by allowing the scored trees to heal. Since the trees do not grow in a uniform manner, and since a tree should be at least six years old prior to its first latex harvesting, this collection process remains a labor intensive one that has not been mechanized.

https://doi.org/10.1515/9783110671094-021

21.2.2 Synthetic rubber

Synthetic rubber always must utilize crude oil as its source raw material. Monomers that are used in various synthetic rubber formulations include: isoprene (aka 2-methyl-1,3-butadience), isobutylene (methylpropene), and chloroprene (2-chloro-1,3-butadiene), although the lattermost monomer requires the introduction of the chlorine atom through a chemical process. The first is essentially the same material that is biosynthesized in the Brazilian rubber tree. The monomer, and a simplified view of its polymerization, is shown in Figure 21.1.

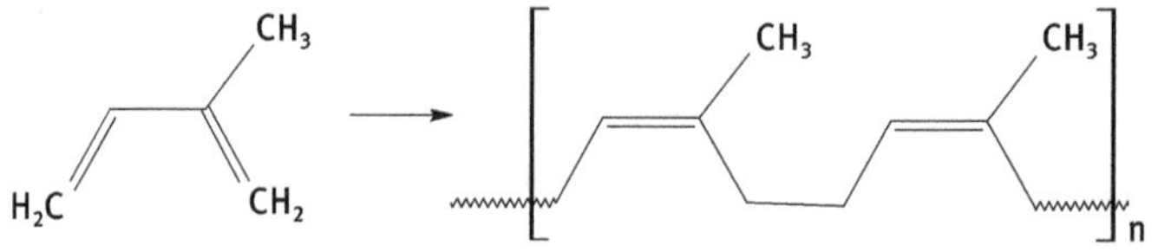

Figure 21.1: Isoprene Polymerization to Rubber.

Polymerization of a single monomer is not an absolute requirement for the production of the various synthetic rubbers. Indeed, co-polymerization of one or more different monomers, as well as the inclusion of small amounts of additives, is often the method of choice to produce rubbers with specific properties – usually chemical inertness, physical toughness, or elasticity in extremes of temperature [6, 4, 5, 1, 2].

21.2.3 Vulcanization

The story of the accidental vulcanization of rubber by Charles Goodyear has become well known among chemists and chemical engineers [9]. The introduction of a small amount of sulfur provides a series of sulfur bridges – cross links between the polymer chains of the isoprene starting material – which lock the chains in place, but not so completely as to force the loss of all stretching and elasticity. Vulcanized rubber is able to return to its starting shape, even after being subjected to extremes of temperature.

While the cross-linking of polymer chains with sulfur is not a process that occurs in a convenient, stoichiometric reaction, the general repeat unit can be easily shown. Figure 21.2 shows a repeat unit of vulcanized rubber. The sulfur connection shown does not have to be at the same position from one chain to another, and sometimes does link back to the same chain. The overall effect of cross-linking though is to create what might be considered a molecular net, whereas unvulcanized rubber would be more properly considered molecular chains.

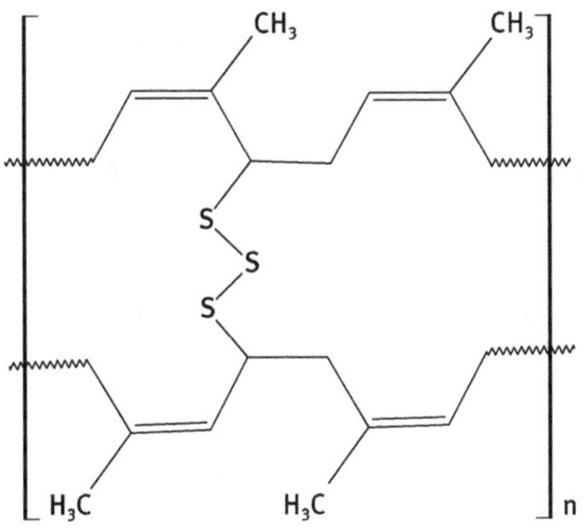

Figure 21.2: Vulcanized Rubber, Schematic.

21.3 Major uses

While rubber tires and tubes are the major end product most people think of when considering the uses of rubber, there is a long list of consumer end products which incorporate rubber. Rubber tires and rubber tubes do currently occupy more than half of the end-use materials made of rubber, but the textile industry uses rubber fibers in the production of garments.

A listing of rubber uses includes: tires, tubes, motion dampeners, matting and flooring, belts and hoses in the automotive industry, textiles, and adhesives. All these can be made from either natural or synthetic rubber, although end users have preferences for one sources versus another in specific applications.

21.3.1 Synthetic isoprene rubber

The rapid growth of synthetic rubber as an industry can be traced to the Second World War, and the United States' need for the material, when all the natural sources in southeast Asia were under the control of the Empire of Japan. For a brief time during the war, the United States government subsidized the construction of factories for the production of synthetic rubber. Evidence of how much the United States needed rubber for the war effort is reflected in two numbers: in 1941 there were 8,000 tons of synthetic rubber produced, and in 1945 that number had increased to 820,000 tons.

21.3.2 Poly-butadiene rubber

Polybutadiene rubber (BR) is currently one of the high volume rubber formulas used in various tires, and specialty applications that require tough plastics, such as golf balls.

The toughness of the material, as well as the low cost of production, makes BR one of the rubber formulations of choice for the near future.

21.4 Other elastomers

Several other starting monomers can be used in producing what has now become a variety of synthetic rubbers. Table 21.1 shows some of the principle ones.

Table 21.1: Synthetic Rubber, Starting Materials.

Name	ISO 1629 Code	Example Use
Epichlorohydrin	ECO	Resins
Fluorinated hydrocarbons	FKM	Viton seals and O-rings
Polybutadiene	BR	Tires
Polychloroprene	CR	Clothing
Polysiloxane	SI	Silicone rubber seals
Polyurethanes	PU	Foam insulation

The ISO 1629 Code is an International Organization for Standardization code used for the classification of types of rubber, so that there is no confusion in international business over names, which might occur if trade names alone were used as identifiers.

21.5 Recycling

The amount of reaction chemistry that is involved with producing finished rubber, and the additives that are used in the vulcanization process, all of which ensures its toughness and durability, argues that there will probably never be large rubber recycling programs. The material is too tough to be used as a feedstock for some process that simply involves melting and re-casting. There are however, relatively small-scale forms of rubber re-use, which usually depend upon the ability to mechanically shred or crush the rubber into small pieces, then bind it into some other form. Figure 21.3, below, shows rubber paving bricks used at the elite horse farms of Kentucky, so that prize thoroughbred race horses do not have to walk on a harder brick surface. Additionally, shredded rubber tires are sometimes used as a surface for outdoor children's playgrounds.

In the recent past, there has been increased interest in both recycling rubber and in producing it in a sustainable manner [3]. This concern is based on the changes in climate that have occurred in the past twenty years, and on how this may affect the growth of rubber trees, especially in Eastern Asia.

Figure 21.3: Interlocking Paving Bricks Made from Recycled Rubber.

Bibliography

[1] Australasian Plastics and Rubber Institute, Inc. Website. (Accessed 7 December 2022, at www.apri.org. au/).
[2] Belgian Plastics and Rubber Institute. Website. (Accessed 7 December 2022, at www.bpri.org/).
[3] Grandin, G. "Fordlandia: The Rise and Fall of Henry Ford's Forgotten Jungle City," Metropolitan Books, 2010.
[4] Indian Rubber Institute. Website. (Accessed 7 December 2022, at www.iri.net.in/).
[5] Indian Rubber Institute, Kerela Branch. Website. (Accessed 7 December 2022, at http://www.iri.net.in/ kerala-branch/).
[6] International Institute of Synthetic Rubber Producers. Website. (Accessed 7 December 2022, at https://www.iso.org/organization/9567.html).
[7] International Rubber Research and Development Board. Website. (Accessed 7 December 2022, at https://www.iso.org/organization/9615.html).
[8] Jackson, J. "The Thief at the End of the World: Rubber, Power, and the Seeds of Empire," Viking, 2008.
[9] Korman, R. "The Goodyear Story: An Inventor's Obsession and the Struggle for a Rubber Monopoly," Encounter Books, 2002.

22 Silicon

22.1 Introduction, purification of silicon

Silicon does not occur in its reduced form naturally, although it is a common element in the Earth's crust. Many sands and rock types are composed largely of silicates, meaning silicon combined with oxygen (indeed, 51 % of the Earth's oxygen is locked in the silicate rocks of its crust). Sand itself has been used for thousands of years in the production of glass.

The extraction of silicon from raw silica is easy to show in a chemical reaction, as seen below:

$$SiO_2 + 2C \longrightarrow Si_{(l)} + 2CO_{(g)}$$

But, as with some other reductions, the isolation of this element requires enormous amounts of energy. The above reaction is normally carried out at 1,900 °C in an elec-

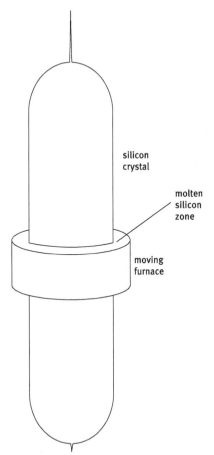

silicon
crystal

molten
silicon
zone

moving
furnace

Figure 22.1: Zone Refining Furnace.

https://doi.org/10.1515/9783110671094-022

tric arc furnace. Carbon electrodes are routinely employed, and the liquid silicon is normally collected at the base of the furnace. Cooling the element solidifies it, and allows its further purification with a zone refining furnace, seen in Figure 22.1. The zone refining furnace is capable of concentrating impurities in the molten zone while it moves slowly downward along a silicon crystal. Since the furnace moves downward, gravity pulls the impurities downward as well. Repeated passes of the furnace along a silicon crystal result in progressively higher purity.

22.1.1 Ferrosilicon

Ferrosilicon is a set of alloys of iron that can be as high as 90 % silicon. It is manufactured for applications where a lightweight metal is useful or required. It is often used where aluminum or aluminum alloys are the main application.

Currently, the largest producer of ferrosilicon is China, as can be seen in Figure 22.2. But internationally, it can be seen that it is produced in several countries.

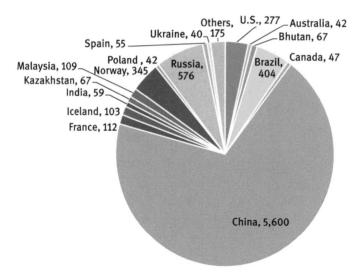

Figure 22.2: Ferrosilicon Use by Country.

The combined total is 8.1 million metric tons (based on 2021 production) [6].

22.2 High purity silicon, uses

Highly pure silicon is required for the computer and electronics industry, and while it does not consume enough silicon to make it one of the top 100 chemicals produced globally each year, much of our modern life and the aspects of it that are now taken for granted would not be achievable without the computers made with such highly pure silicon. The purity level in these applications is referred to as "nine-nines," meaning 99.9999999 % pure.

The Czochralski Process is currently the least expensive method to obtain such high purity. This process involves inserting a precisely oriented seed crystal of silicon into molten silicon, in an inert atmosphere, using a quartz crucible, then slowly pulling the seed crystal upwards while the rod supporting the seed is rotated. The process was first developed in the Bell Labs, in the search for high purity material for transistors. The process has similarities to the zone refining furnace just mentioned.

22.3 Silicones

One of the major uses of silicon-containing materials is the now vast area of silicones. These polymers are a mixture of organic and inorganic components, and incorporate a repeating silicon – oxygen bond as the backbone. The general structure for the repeat unit of a silicone is $[R_2SiO]_n$, where R is some organic functional group, such as a methyl (although it can be many other functional groups, including bi-functional, cross-linking organic groups).

The properties of silicones are often directly a result of which side group is pendant to the main chain, with various groups producing liquids, and others forming rigid gels. This level of control of the properties of the resulting material is extremely useful when designing polymers for specific end user applications. This also positions what can be called the silicone industry on something of a fulcrum between large, industrial scale chemistry, and smaller, specialty syntheses.

22.3.1 Production of silicones

Silicones are produced from refined silicon, an organo-chlorine molecule, and reduced copper. While the reaction chemistry can be simplified to that shown in Scheme 22.1, there are many variations, based on the organic starting material. In most, the organo-chlorine molecule is reacted with some fluidized bed of silicon. Up to 10 % copper must be present and plays a catalytic role.

The production of organo-chlorine molecules has already been mentioned in Chapters 6 and 10. In the case of methyl chloride, shown in Scheme 22.1, it is formed as follows:

$$CH_3OH + HCl \longrightarrow H_2O + CH_3Cl$$

Both methanol and hydrochloric acid are inexpensive starting materials.

$$2CH_3Cl + 2Cu + Si \longrightarrow (CH_3)_2SiCl_2 + 2Cu$$

Including as a non-isolable intermediate:

$$CH_3Cl + 2Cu \longrightarrow CH_3Cu + CuCl$$

As well as:

$$CuCl + Si \longrightarrow SiCl + Cu$$

Scheme 22.1: Chemistry of the Production of Silicone [3].

From the production of the monomer, production of the silicone polymers can be represented by the relatively simple reaction:

$$nSi(CH_3)_2Cl_2 + nH_2O \longrightarrow [(CH_3)_2SiO]_n + 2nHCl$$

This reaction represents the synthesis of one of the most common silicones, polydimethylsiloxane, or PDMS, which exists as an oil. But it also serves as an example for the synthesis of numerous other silicones as well [1].

22.3.2 Uses and features of silicones

The field of silicones has become extremely broad precisely because of the ease with which end products can be tailored to specific uses and material performances. The general synthetic scheme shown in Scheme 22.1 must be modified in terms of reaction conditions to produce silicones with different side chains; but this is not a difficult task.

Silicone producers usually divide their product offerings by types of application. For example, the Silicone Industry Association of Japan [4] lists the following:
1. Construction
   - Protective coatings
   - Sealants
   - Others
2. Electronics
   - Electronics
   - Appliances
   - Power and cables
   - Others
3. Industrial uses
   - Industrial and mold making
   - Plastics, chemicals, industrial use additives
   - Others

4. Personal care and life style
   - Health care and medical care
   - Personal care and home care
   - Others
5. Transportation
   - Automotives
   - Others
6. Special systems
   - Adhesives and coatings
   - Papers and films
   - Textile and leather
   - Others

In the United States, Dow Corning, which is a major producer of silicones, boasts an even greater diversity of products and applications utilizing one silicone material or another [1]. Clearly, the list of consumer materials employing silicones is extensive [1, 4, 5, 2].

22.4 Silicon dioxide

Silicon dioxide, or silica, is not only used for the production of elemental silicon, as discussed at the beginning of this chapter, but is also used as the main material for the manufacture of glass and much fiber optical cabling. The production of pure silicon dioxide for specialty uses is more complex than simply mining and cleansing silicate sand of its macroscopic impurities, but such needs remain small on the industrial scale. Most silicon dioxide forms glass according to the broad reaction:

$$\left(\frac{1}{4} - \frac{3}{4}\right)Na_2O + SiO_2 \longrightarrow (glass)$$

If the stoichiometry is altered to a 1:1 ratio, the result is sodium silicate, as follows:

$$Na_2O + SiO_2 \longrightarrow Na_2SiO_3$$

Strictly speaking, for the production of glass, other metal oxides beside sodium oxide can be used. But sodium oxide remains the most common additive in glass manufacturing. Silicon and silicon dioxide are tracked by the USGS, and considered common materials [6].

22.5 Recycling

22.5.1 Recycling of silicon

The recycling of silicon is usually meant to indicate the recycling of end products, such as computers. Curiously, there are more valuable components in computers than silicon, meaning the other, rare, refined metals. These metals are recovered and recycled, while the silicon chips are not.

The silicon used in ferrosilicon alloys is not recycled as an element, and is noted as such in the USGS Mineral Commodities Summary each year. It further mentions that because silicon is so common in silicates that it is estimated there will be ample supplies for decades [6].

22.5.2 Recycling of silicones

Recycling of silicones is difficult because of the wide variety of these materials. Additionally, many types of silicones are produced for consumer end-use products, including the large number of personal care products on the market. These are made for a single use and disposal. Although breakdown of silicones in landfills is extraordinarily slow, there have been no reports that their presence in landfills is a health risk for humans [1].

22.5.3 Recycling of silica

The recycling of glass, particularly post-consumer use glass containers, has become an established practice in large parts of the world. The reasons are both economic and environmental. Economically, it is much less expensive to produce glass from refined, crushed glass than it is to manufacture it from virgin silica sources. Environmentally, glass simply does not break down in landfills (indeed, glass objects are an entire sub-discipline within archaeology, just because they do not breakdown even over millennia). Thus it is environmentally friendlier to re-use glass material and objects. In many areas, glass bottles are not even crushed and re-melted. When distributors, such as grocery stores, reclaim usable glass bottles, they can be returned to the manufacturer and bottler for disinfection and re-use.

Bibliography

[1] Dow Corning. Website. (Accessed 10 December 2022, at www.dow.com/).
[2] European Silicones Center. Website. (Accessed 10 December 2022, at www.silicones.eu/).
[3] Greenwood, N. N. and Earnshaw, A. "Chemistry of the Elements," Pergamon Press, 1984, pp. 421–424.
[4] Silicone Industry Association of Japan. Website. (Accessed 10 December 2022, at www.siaj.jp/en/).

[5] Silicon Manufacturers Group, SEMI. Website. (Accessed 10 December 2022, at www.semi.org/en/ About).

[6] USGS Mineral Commodity Summaries 2022, US Department of the Interior, U.S. Geological Survey, https://doi.org/10.3133/mcs2022.

23 Iron and steel

23.1 Introduction and historical production

Perhaps no single element has helped define human civilizations more than iron. Iron ages tend to come after Bronze ages as localized civilizations develop. This appears to be so simply because the melting point of iron is higher than that of different bronze alloys. Iron tools wear more slowly than those made of bronze or other metals, and because iron weapons last longer in battle, societies with them tend to be able to subdue, incorporate, or exterminate societies that do not have them. Some of the earliest iron artifacts found in ancient Egyptian tombs are small swords made from meteoric iron, which were apparently given the rather romantic name, "daggers from heaven."

But the greatest phenomenon that people have found historically, that is connected only to iron, is the ability to pull it from the fires and forges in which it is refined as a metal that has almost magical properties to it – steel.

For most of history, the production of steel was a small scale operation, one that produced one tool or weapon at a time. The magic involved in it is the result of most smiths not consciously knowing that a certain amount of carbon had to be incorporated into the iron matrix, and that this occurred in the process of repeated heating and cooling in a coal-fed fire. Thus, the coal served as the carbon source that produced steel under just the right circumstances.

By the Middle Ages, iron production had become large enough that it was a separate industry, and in 1855 Henry Bessemer advanced its production through the use of what is now called the Bessemer converter. This forced oxygen through the molten iron, and thus removed a significant amount of impurities from each molten batch. These converters were large enough that some could produce up to 30 tons of iron at a time.

23.2 Ore sources

Iron ores are found and refined on all six inhabited continents. Both hematite and taconite ores have been used for the iron they contain. Prior to the Second World War, slag heaps in Italy from iron refining during the time of the Roman Empire were re-used and re-refined, because the Roman working of iron, while voluminous for its time, was inefficient.

Figure 23.1 shows pig iron production for 2011, while Figure 23.2 shows raw steel production for the same year, both in millions of metric tons [9].

https://doi.org/10.1515/9783110671094-023

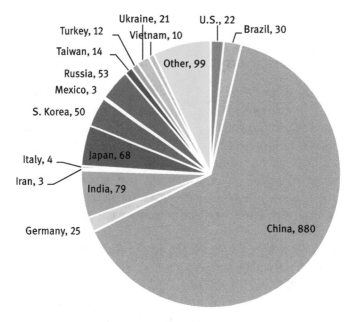

Figure 23.1: Pig iron production, in millions of metric tons.

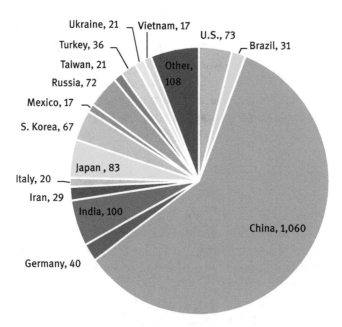

Figure 23.2: Raw steel production, in millions of metric tons.

23.3 Current iron production

The Industrial Revolution accelerated the production of iron from a level that was akin to a cottage industry up to the large scale process we know today. Numerous applications for iron made it omnipresent in the developing world, from buildings, to bridges, to weapons. Some companies that are not at all associated with the production of armaments at least for a time had an interest in iron weapons production. An example is the Ford Motor Company, one of the world's best known car companies, at least for a time had the ability to make cannons, shown in Figure 23.3, below.

Figure 23.3: Iron cannon with Ford logo, at right.

But the large scale production of iron today goes far beyond automobiles and weaponry. Iron and steel alloys are used in an enormous number of applications, from small, household items, to structural support beams in skyscrapers, and in some cases for bridges.

23.3.1 Blast furnace

The main means by which iron ores are reduced to iron metal is through a blast furnace, seen in Figure 23.4. While such furnaces can still be constructed by a few individuals, for use in making several hundred pounds of iron, perhaps for artisanal purposes, industrial blast furnaces are taller than many buildings, and the decisions to build, operate, and decommission them are corporate ones.

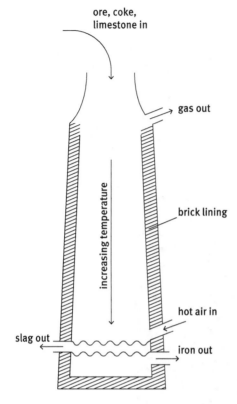

ore, coke, limestone in

gas out

increasing temperature

brick lining

hot air in

slag out

iron out

Figure 23.4: Blast Furnace.

The reaction chemistry of a blast furnace can be shown very simply in a single reaction:

$$Fe_2O_3(s) + 3CO(g) \longrightarrow 2Fe(l) + 3CO_2(g)$$

Since there is certainly more to the process, this can be displayed in more detail, with a breakdown of the reduction steps listing in correspondence with the temperature zone at which a particular reduction occurs.

$2C(s) + O_2(g) \longrightarrow 2CO(g)$	200–700 °C
$3Fe_2O_3(s) + CO(g) \longrightarrow CO_2(g) + 2Fe_3O_4(s)$	600–700 °C
$Fe_3O_4(s) + CO(g) \longrightarrow CO_2(g) + 3FeO(s)$	850–900 °C
$CaCO_3(s) \longrightarrow CaO(s) + CO_2(g)$	850–900 °C
$FeO(s) + CO(g) \longrightarrow Fe(l) + CO_2(g)$	1000–1200 °C
$C + CO_2(g) \longrightarrow 2CO$	1300 °C
$SiO_2(s) + CaO(s) \longrightarrow CaSiO_3(l)$	

Limestone, discussed in Chapter 5, as well as coke, are used in large quantities to provide a flux and reducing agent, in order to capture any impurities in the original ore, and to reduce the iron oxides to elemental iron.

23.4 Steel production

Iron may be the only element with any industrial use that actually has enhanced properties when it is properly alloyed with small amounts of impurities. Pure iron is deformable, and does not keep its shape under extreme stress, while steel – meaning iron with 1–4 % carbon alloyed into it – is harder and retains its shape under a variety of extreme conditions.

A large number of steel alloys exist, incorporating small amounts of other metallic elements, as well as varying amounts of carbon. There are a number of trade organizations in different countries that are involved in the production and regulation of steel in its many forms [1, 10, 5, 3, 2, 6, 7, 4, 8]. The production of so many alloys has been the decades long work of metallurgists, chemists, and chemical engineers looking for specific properties within an iron-based material. Table 23.1 illustrates some of the more common types of steel, although there are many others.

Table 23.1: Common Types of Steel.

Type of Steel	Alloying Element(s)	Use or Property
Carbon steel	C	Hardness
Stainless steel	Cr 11 %, Ni	Corrosion inhibition
High speed steel	W	High hardness
High strength low alloy steel	Mn, 1.5 %	Greater strength
Low alloy	Mn, Cr, <10 %	Increased hardness
Tool steel	W, Co	Increased hardness, drills, cutting tools
Mn steel	Mn, 12 %	Durability

23.5 By-products

Unfortunately, the refining of iron from its ores, and the production of steel, produce significant amounts of by-products and waste. Using the reactions listed above, one can see that CO_2 is given off at several steps, and that slag is co-produced with iron. Slag is less dense than iron, and thus is poured off at a point above the outlet for the liquid iron. Slag is sometimes used in the production of Portland cement.

One by-product that does not appear in the reaction chemistry shown above is sulfur or sulfur oxides. Iron ores are not usually 100 % iron oxides. Mixed oxide-sulfide ores produce sulfur-containing by-products. When this occurs in the gas phase, SO_2 and SO_3 can be produced and emitted.

Overall, for every ton of iron produced, significantly more than one ton of carbon dioxide, slag and other waste products are produced, and in the case of carbon dioxide, emitted to the atmosphere.

23.6 Recycling

Few materials are recycled as often as iron and steel. Scrap yards exist as a large, diverse industry in both developed and developing countries. Most scrap yards and recycling centers do not focus exclusively on a single metal, but almost all will deal with old iron or steel materials. Simple tests, such as running a magnet over the sample, lets scrap yard operators know if the grade of iron is stainless steel or another grade (non-magnetic versus magnetic). The primary source of iron and steel scrap within the United States remains the recycling of automobiles [9]. The recycling of steel has grown large enough that a trade organization, the Steel Recycling Institute, is organized around it [1, 8]. As might be expected, the organization is keenly aware of, and advertises, how much energy is saved by recycling steel, as opposed to producing the alloy from ores.

Currently, efforts are underway to examine how to reduce the amount of CO_2 emitted to the atmosphere during iron refining. Since there is no model to follow, at least not at the industrial level, several possibilities are being examined, including any form of carbon dioxide sequestration, the possibility of using a different reducing agent, or the possibility of using a biologically renewable reducing agent. None have yet been scaled up to industrial production, however.

Bibliography

[1]   American Iron and Steel Institute. Website. (Accessed 10 December 2022, at www.steel.org/).
[2]   Australian Steel Institute. Website. (Accessed 10 December 2022, at steel.org.au/).
[3]   British Stainless Steel Association. Website. (Accessed 10 December 2022, at www.bssa.org.uk/).
[4]   International Nickel Study Group. Website. (Accessed 10 December 2022, at www.insg.org/).
[5]   International Stainless Steel Forum. Website. (Accessed 10 December 2022, at worldstainless.org/).
[6]   Japan Iron and Steel Federation. Website. (Accessed 10 December 2022, at www.jisf.or.jp/en/).
[7]   South African Iron and Steel Institute. Website. (Accessed 10 December 2022, at www.saisi.co.za/).
[8]   Steel Recycling Institute. Website. (Accessed 10 December 2022, at www.steel.org/recycling).
[9]   USGS Mineral Commodity Summaries 2022, US Department of the Interior, U. S. Geological Survey, https://doi.org/10.3133/mcs2022.
[10] World Steel Association. Website. (Accessed 10 December 2022, at www.worldsteel.org/).

24 Aluminum

24.1 Introduction and history

Aluminum is one of the "newer" metals, isolated only in 1827 by Friedrich Woehler, despite it being estimated to be the most abundant metallic element within the Earth's crust. It always occurs in nature bound as an ore, usually bauxite.

For a few decades after its discovery, aluminum was an extremely expensive metal, because the extraction process was energy intensive and the materials for it were themselves rare. The first breakthrough in aluminum refining came in 1859, with the introduction of the Deville Process for alumina production. This made aluminum more accessible, and significantly less expensive. It was however the introduction both of cheap electricity in the 1880s and the Hall–Heroult Process which made this newer process a viable one, and sent the price of aluminum plummeting. Since the patenting of this process in 1887, the price of aluminum has dropped to the general area in which it remains today. Also, aluminum is produced in the same manner today that it has been since the patenting of the Hall–Heroult Process.

Figure 24.1 shows the decline in price for aluminum over the past decades. This is one of only a few metals for which such a graph can be made. While many metal elements were discovered in the 1800s, very few have become as industrially useful as aluminum, and thus very few have had their value tracked as closely.

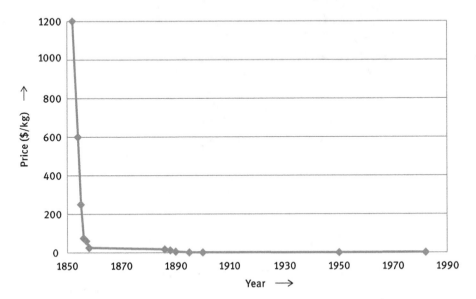

Figure 24.1: Aluminum Prices From 1827 [7].

https://doi.org/10.1515/9783110671094-024

The largest aluminum manufacturers globally today are Rio Tinto Alcan, headquartered in Montreal, Canada [8], Rusal, headquartered in Moscow, Russia (with financial headquarters in Jersey) [9], and Alcoa, headquartered in Pittsburgh, in the United States [1]. Alcoa has corporate interests in over thirty countries.

24.2 Bauxite sources

Bauxite is not a particularly rare mineral, but extracting it from its known sources can be labor intensive. Large deposits are located in China and Canada. Figure 24.2 shows aluminum production worldwide, as a total of 44.1 million metric tons, produced in 2011 [10]. Virtually all this aluminum comes from bauxite, although there are other aluminum-containing ores (none of which are currently economically competitive with bauxite for exploitation) [7].

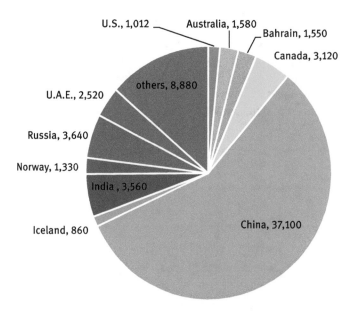

Figure 24.2: Aluminum production worldwide, in thousands of metric tons.

24.3 Production methods

All aluminum refined in the world today is produced using the Hall–Heroult Process. This is an electrolytic process that requires molten cryolite (Na_3AlF_6), which brings down the working melting point of alumina from 2000 °C to approximately 1000 °C. A small amount of aluminum fluoride (5–10 %) is also added to what is called the cell to further lower the melting temperature of the mixture.

A diagram of a Hall–Heroult Process cell is shown in Figure 24.3. All cells have a brick ceramic wall, usually encased in a steel shell or container. The two electrodes are both carbon, and the anodes must be replaced to keep the direct current running through the cell. This loss of the electrode over time means the evolution of CO_2 which must be captured and vented. In addition, some HF is generated, which must also be captured.

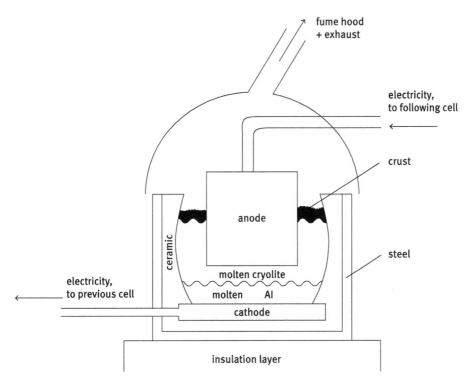

Figure 24.3: Hall–Heroult Cell.

Liquid aluminum is more dense than liquid cryolite (the reverse of their densities as solids), which is why it sinks to the bottom of the cell. This must then be vacuum siphoned out of the cell for casting into ingots. Fresh alumina must be added to the top of cells as the electrolysis continues.

Because of the large amount of electricity required for aluminum refining, smelters are usually located near inexpensive sources of electrical power. In many cases, aluminum smelters are located near dams, because of the inexpensive hydroelectric power. In some cases, power is directed to the smelter site. And in one case globally, raw material is shipped to the power source – Iceland. Because of this small nation's abundant geothermal energy, it has been profitable to refine aluminum by bringing the alumina and cryolite to the power source.

While the reaction chemistry of the Hall–Heroult Process can be a complex mixture of aluminum fluorides and oxofluorides, the cathode and anode reactions can be simplified to the following oxidation and reduction:

$$2O^{2-}(l) + C(s) \longrightarrow CO_2(g) + 4e^-$$
$$Al^{3+}(l) + 3e^- \longrightarrow Al(l)$$

24.4 Major industrial uses

Many people consider the major use of aluminum to be airplane parts and beverage cans, but there is a wide variety of other consumer uses for the metal. In most cases, aluminum is chosen as a desired material because of its extremely low density (2.70 g/cc) or its low reactivity with its surroundings.

The percentages for particular uses change annually, but the following, shown in Table 24.1, have been the major uses of aluminum over the long term.

Table 24.1: Uses for Aluminum.

Use	Example
Machinery	Heat exchangers
	Chemical equipment
Building and construction	Building panels
	Windows
	Doors
	Mobile home parts
Containers	Cans
	Foil
	Chemical containers
Transportation, parts for:	Aircraft
	Cars
	Light trucks
	Trailers
	Shipping containers
Electricity	Power lines
	Tower and structural supports
Consumer goods	Refrigerators
	Kitchen appliances
	Air conditioning equipment
Other	Recreational boats
	Camp gear
	Utensils

24.5 By-products

As with the production of other metals that must be reduced from an oxide or sulfide ore, the reduction of aluminum to the metal generates a significant amount of by-products. A ton of aluminum consumes almost 1.9 tons of Al_2O_3, which means the oxygen is captured by the anodic carbon as CO_2. The carbon for this is provided by almost $\frac{1}{2}$ ton of anode material. Multiplying these numbers by the 44 million metric tons of aluminum produced in 2011, as described in Figure 24.2, gives an indication of how much CO_2 was generated.

As mentioned, HF is a further by-product of aluminum smelting, and must be much more carefully controlled than CO_2. While carbon dioxide can be released into the atmosphere, there are immediate, often severe, consequences should an accidental escape of HF occur.

24.6 Recycling

Aluminum recycling is a mature sub-section of the scrap metal recycling industry [2, 6, 5, 3, 4]. It has been noted that it saves 95 % of the electricity that would be used when compared to producing aluminum from ore [2, 6, 5]. Most individuals consider aluminum cans the prime type of aluminum that is recycled; and in truth aluminum cans are now recycled in the US at a rate of approximately 65 % [2].

But there are many other aluminum objects and grades that are recycled as well. Aluminum light towers and building materials weigh enough that they are almost always financially worthwhile to recycle. Indeed, when the price of aluminum rises, there have been numerous cases in cities worldwide where aluminum street lights have been stolen and turned in to scrap yards for the value of the metal. Aluminum from building demolition is routinely recycled.

Bibliography

[1] Alcoa. Website. (Accessed 10 December 2022, at www.alcoa.com/global/en/home).
[2] The Aluminum Association. Website. (Accessed 10 December 2022, at www.aluminum.org/).
[3] Australian Aluminum Council. Website. (Accessed 10 December 2022, at aluminium.org.au/).
[4] Aluminum Federation South Africa. Website. (Accessed 10 December 2022, at www.afsa.org.za/).
[5] European Aluminum. Website. (Accessed 10 December 2022, at European-aluminum.eu).
[6] Onestopalu.dk. Website. (Accessed 10 December 2022, at onestopalu.dk).
[7] Price history for aluminum. Website. (Accessed 10 December 2022, at www.indexmundi.com/commodities/?commodity=aluminum).
[8] Rio Tinto Alcan. Website. (Accessed 10 December 2022, at www.riotinto.com/).
[9] Rusal. Website. (Accessed 10 December 2022, at www.rusal.ru/en/).
[10] USGS Mineral Commodity Summaries 2022, US Department of the Interior, U. S. Geological Survey, https://doi.org/10.3133/mcs2022.

25 Copper

25.1 History

Copper has played a role in the rise of numerous civilizations. The ability to smelt copper, and the ability to make two alloys of it – brass and bronze – has led to improved tools and weapons for people throughout history. Indeed, the use of bronze in a society often leads to that time period being called "the Bronze Age."

In the west, the word 'copper' has its roots in the name of the island of Cyprus, where extensive copper deposits were found and mined in ancient times.

In ancient China, copper was alloyed with tin more than two thousand years ago, resulting in cast bronze items that were made for functional applications as well as ceremonial use. Han Dynasty bronze objects have become valuable items on the world's art markets today.

There is also archaeological evidence that pre-Columbian native peoples of North America may have traded copper from as far north as the modern-day US – Canadian border down as far as present-day Central America.

25.2 Ore sources

Copper ore is found on all six inhabited continents, and may also be available in Antarctica (although current treaties prevent mining on this land mass). Copper is found as oxide and sulfide ores, and as placer copper, or reduced metal. This is a material of interest to the United States Geological Survey, and mine production as well as scrap recovery and recycling are tracked [9]. In 2011, the United States produced 1.1 million tons of copper through mining, and 130,000 tons from recovery and recycling of old scrap [9]. The Copper Development Organization also tracks the recovery of copper scrap and its applications [6].

25.3 Producers

The number of copper producers is very large, with members of the American Copper Council [1], and the Copper Alliance [5] together totaling more than 100 companies on six continents. These and other trade organizations exist to share information concerning techniques on copper refining and production.

25.4 Production methods

While placer copper has been found in numerous locations throughout the world, copper is routinely mined. The largest single copper nugget is believed to be the 420 ton

https://doi.org/10.1515/9783110671094-025

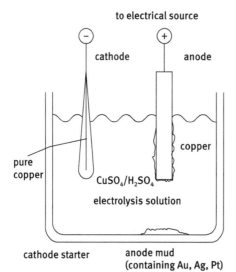

to electrical source

cathode   anode

copper

pure copper

$CuSO_4/H_2SO_4$

electrolysis solution

cathode starter   anode mud (containing Au, Ag, Pt)

Figure 25.1: Electrolytic Refining of Copper to High Purity (>99.9 %).

piece currently located in the upper peninsula of Michigan. Mining usually involves the following series of steps, with some differences depending upon the batch of ore that is the starting material for the process, specifically whether it is predominantly an oxide versus a sulfide ore:

1. Mining and crushing. Most copper is mined as either oxide ores or sulfide ores, although copper metal can also be mined. After extraction from the ground the ore must be crushed into workable pieces, generally 1–2 cm in size.
2. Grinding. This step further crushes ores in large rotating machines, until it has a powder consistency. The copper concentration at this stage is quite low (ca. 1 %).
3. Concentration. Using water, the powders are concentrated to >10 %, and waste products are extracted and removed.
4. Leaching. A copper sulfate solution is produced by leaching oxide ores with a relatively weak acid solution.
5. Electrolytic processing. (electrowinning) Pure copper cathodes are inserted into a process tank, which is electrically charged, to allow the movement of the copper ions to the electrodes. This process builds up extremely pure copper at the cathodes, and also can produce an anode mud, as seen in Figure 25.1. The anode mud is collected and, if necessary, refined to extract silver, gold, or other precious metals that were part of the predominantly copper-containing solution.

Or
4. Smelting. Smelting is the process of melting and purifying the metal, usually in multiple steps, which can result in up to 99 % pure copper.
5. Electrolytic Refinement. Much like electrowinning, this refinement technique allows copper ions to deposit at pure copper cathodes, ultimately gaining significant

mass (up to 300 lb. cathodes can be formed). Again, as with electrowinning, other precious metals can be recovered at this step.

6.   Copper cathode conversion.

At this point, copper cathodes can be sold to users, and formed into the end products, such as wiring, piping, and sheet, used by numerous companies.

25.5 Major industrial uses

25.5.1 Wire, piping, machinery, and alloys

Copper wiring and piping remain major end uses for the metal [7, 4, 3, 2]. Wiring is produced in numerous thicknesses, referred to as gages, and is used in virtually all residential devices utilizing electricity, as well as an enormous number of industrial and commercial devices.

Copper piping is also used in a very large number of applications, from household plumbing to industrial operations that require large volume fluid flows.

25.5.2 Coinage

Copper coinage has been a continued use for copper for millennia. The volume of it produced on a global scale in the past thirty years is significantly larger though than it has been in the past. Additionally, in the nineteenth and twentieth centuries, as coinage assays were improved and the weight and fineness of coins in developed nations were standardized, copper was also used as the alloying element in silver and gold coins. The term "coin silver" usually means silver metal with 10 % copper in it, as opposed to sterling silver, which is 92.5 % silver, and the remainder copper. In a similar manner, copper is alloyed with gold. The end result is to make gold coins that have superior hardness and greater durability than would be seen with pure gold coins. For example, the British sovereign, a gold coin still used as a store of wealth today, weighs 7.988052 grams, which is composed of 7.322381 grams of gold and 0.666139 grams of copper. The United States double eagle or $20 gold piece has a somewhat simpler alloy, being 90 % gold and 10 % copper.

25.6 Brass

In general, the terms "brass" and "bronze" mean respectively: copper alloyed with zinc, and copper alloyed with tin. Prior to the 1800s, it is unclear whether or not metallurgists always recognized the difference between the two alloying elements, as many artifacts contain both zinc and tin.

Modern industry uses a variety of different compositions of brass. A partial, alphabetical listing is shown in Table 25.1.

Table 25.1: Compositions of Various Brasses, in Percent (empty boxes indicate the element is absent from a specific brass composition).

Name	Cu	Zn	Sn	Fe	Mn	Ni	Al	Pb	Use
Admiralty brass	69	30	1						Various
Aich's alloy	60.66	36.58	1.02	1.74					In seawater environments
Alpha brass	65	<35							
Prince's metal	75	25							
Beta brass	50–55	45–50							Can be cast
Cartridge brass	70	30							Ammunition casings
Gilding metal	95	5							Ammunition
High brass	65	35							Rivets and screws
Low brass	80	20							Metal adapters
Mn-brass	70	29			1.3				US dollar coins
Muntz metal	60	40		<1					In seawater environments
Ni-brass	70	24.5				5.5			1 pound coins in UK
Nordic gold	89	5	1				5		10, 20, and 50 Euro cent coins
Red brass	85	5	5					5	
Rivet brass	63	37							
Tombac	85	15							Jewelry
Yellow brass	67	33							

Clearly, there are numerous brass formulations. Each has been made for a specific purpose, but often has been found useful for other applications beyond that of the original design.

25.7 Bronze

Bronze currently does not have as many industrial applications as brass, but remains the material of choice for metal sculptors [3], and for large bells. Often, sculptures are simply called "bronzes." Historically, artillery guns have been made from bronze, although steel replaced it as a material as the lethality of warfare increased from the nineteenth into the twentieth century.

25.8 Recycling

Copper mining and manufacturing produces enough waste that recycling has become an economically and environmentally friendly option [2, 8, 10, 11]. Even states or areas without recycling laws have means by which copper can be recycled – usually scrap yards.

The USGS tracks the amount of copper scrap recycled each year [9], although this remains less than 10 % of copper use.

Bibliography

[1]   American Copper Council. Website. (Accessed 10 December 2022, at www.americancopper.org/).
[2]   Bronze.Net. Website. (Accessed 10 December 2022, at www.bronze.net).
[3]   Concast Metal Products Co. Website. (Accessed 10 December 2022, at www.concast.com/links.php).
[4]   Copper – Brass Servicenter Association Modernmetals.com. Website. (Accessed 10 December 2022, at https://www.modernmetals.com).
[5]   Copper Alliance. Website. (Accessed 10 December 2022, at copperalliance.org/).
[6]   Copper Development Association, Inc. Website. (Accessed 10 December 2022, at www.copper.org/).
[7]   International Copper Study Group. Website. (Accessed 10 December 2022, at www.icsg.org/).
[8]   International Lead and Zinc Study Group. Website. (Accessed 10 December 2022, at www.ilzsg.org/static/home.aspx).
[9]   USGS Mineral Commodity Summaries 2022, US Department of the Interior, U. S. Geological Survey, https://doi.org/10.3133/mcs2022.
[10] American Exploration and Mining Association (formerly: Northwest Mining Association). Website. (Accessed 10 December 2022, at https://www.tsnn.com).
[11] Pennies from Hell: In Montana, the bill for America's copper comes due. Website. (Accessed 10 December 2022, at harpers.org/archive/1996/10/pennies-from-hell/).

26 Other major metals for industrial use

26.1 Titanium

Titanium has not been known throughout history, having been discovered only in 1791. Today it is valued as an elemental metal both for its hardness and chemical corrosion resistance, as well as its low density, 4.506 g/cc. It also alloys well with other light metals, and is used in applications where overall weight is of concern.

26.1.1 Sources

Two ores, rutile (TiO_2) and ilmenite (usually written $FeTiO_3$, although sometimes expressed (Fe, Mg, Mn, Ti)O_3 depending upon the source) are the main commercial sources for titanium, although small amounts of it are often incorporated into a wide array of minerals. The two extractive processes used to purify titanium ores are the Kroll Process and the Hunter Process, both discussed below. Commercially important titanium compounds are titanium dioxide (sometimes called titanium white) which was discussed in Chapter 6, titanium tetrachloride, and titanium trichloride. As the name implies, titanium white is often used in pigments, including those used in certain foods, because of its very flat, white look. Titanium tetrachloride is used as a catalyst. Titanium trichloride is a catalyst used in the production of polypropylene.

26.1.2 Worldwide production

Titanium is produced in fewer countries than several of the other elemental metals we have examined in other chapters of the book, simply because it is more localized in the Earth's crust. Figure 26.1 shows titanium production worldwide for 2011, with a total of 186 million metric tons, omitting production from the United States. The USGS Mineral Commodity Summaries notes that data for the United States is withheld to protect corporate propriety information [2, 6].

Two mines have provided a large portion of the world's ilmenite in the recent past. They are:
1. Tellnes Mine, in Sokndal, Norway.
2. Lac Tinto Mine, of the Rio Tinto Group, near Quebec, Canada.

There are other areas which produce significant amounts of ilmenite as well, from ilmenite sands. They include:
1. Richards Bay Minerals, South Africa.
2. Moma mine, owned by Kenmare Resources, Mozambique.
3. Murray Basin, and Eneabba, owned by Iluka Resources, in Australia.

https://doi.org/10.1515/9783110671094-026

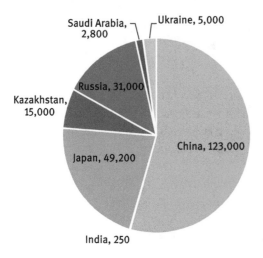

Figure 26.1: Titanium Production, in metric tons.

4. Indian Rare Earth Mineral Mines, Kerala, India.
5. Grande Cote mine, owned by TiZir, Ltd., Senegal.
6. QIT Madagascar Minerals mine, owned by the Rio Tinto Group.

Ilmenite has also been found in lunar rock, and indeed may one day become an economic driver for a return of humans to the moon. The potential to refine titanium by some modification of existing processes, in the air-free environment of the lunar surface, does hold some promise. To date however, no commercial mining concern has announced plans to pursue this.

26.1.3 Extraction chemistry, reactions

Kroll process
The reaction chemistry of the Kroll process can be represented as follows:

$$TiO_2(s) + C \longrightarrow 1,000\,°C \longrightarrow Ti(s) + CO_2(g)$$

which is immediately captured by:

$$Ti(s) + Cl_2(g) \longrightarrow TiCl_4(g)$$

which must then be fractionally distilled to remove other volatile chlorides.
Then:

$$TiCl_4(g) + 2Mg(l) \longrightarrow 2MgCl_2(l) + Ti(s) \quad \text{at } 800–850\,°C$$

The final reaction is performed in stainless steel reactors, which both ensures full reduction of the titanium, and serves as a corrosion-resistant container.

Hunter process

While the Hunter Process has been largely displaced by the Kroll Process for economic reasons, we list it here because it remains an effective way to produce titanium in 99.9 % purity.

$$TiO_{2(s)} + 2Cl_{2(g)} + C_{(s)} \longrightarrow CO_{2(g)} + TiCl_{4(l)}$$

$$TiCl_{4(l)} + 4Na_{(l)} \longrightarrow 700\text{–}800\,°C \longrightarrow 4NaCl_{(l)} + Ti_{(s)}$$

In the first reaction, the carbon source is coke, which consumes the oxygen from the titanium ore. The final reaction is performed in a steel bomb reactor to withstand the heat of the reaction.

26.1.4 Volume produced annually

The amount of titanium produced annually does not completely reflect the size of this industry, because much of the mined ore is not refined to the metal, being refined rather to titanium white or titanium tetrachloride. The United States Geological Survey's annual Mineral Commodity Summaries tracks both the production of titanium sponge and titanium dioxide, and while it does so in units of metric tons, it excludes domestic production tonnage to protect proprietary production methods [6]. Recent import figures in percentage are summarized in Table 26.1.

Table 26.1: Ti and TiO_2 Imported to the United States, 2012.

Country	Ti sponge (%)	TiO$_2$ (%)
Canada		41
China	5	13
Germany		6
Finland		6
Japan	37	
Kazakhstan	51	
Russia	4	
Other	3	34

26.1.5 Major uses

The International Titanium Association (ITA) lists close to fifty major end uses for titanium, including everything from aircraft to golf clubs to ultracentrifuges [2]. Most educated people however tend to associate the use of titanium metal with the aircraft and aerospace industry, because of its already-mentioned toughness and low density. Bearing this out, the ITA website states:

"The commercial aerospace industry is the single largest market for titanium products primarily due to the exceptional strength to weight ratio, elevated temperature performance and corrosion resistance." [2]

It is also well known that titanium components are used in a variety of military hardware applications.

Curiously, ruthenium is at times added to titanium to make an even more corrosion resistant alloy than the titanium alone. Even small amounts of ruthenium, such as 0.1 %, tend to increase the alloys resistance as much as 100 times. This is a significant use of a metal, ruthenium, which has no other current industrial use.

26.2 Chromium

26.2.1 Sources

Chromium ores are mined in several countries throughout the world. Figure 26.2 shows them as a percentage, the whole being approximately 24 million metric tons, excluding the United States for reasons of proprietary production methods (US reserves estimates are 620,000 metric tons) [6]. As with many ores, all chromium-containing ores do not have the same percentage of chromium within them, and thus the extractive technologies must at times be adapted to the ore batch.

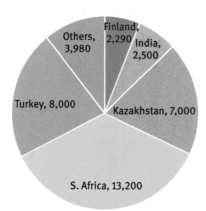

Figure 26.2: Chromium Production by Country, in thousands of metric tons.

26.2.2 Extraction chemistry, reactions

The extraction of chromium from its ores is not as straightforward as that for metals such as iron or lead, because most chromium ores such as chromite also have iron in them. Iron is removed at elevated temperatures because the ore is heated with a sodium carbonate and calcium carbonate mixture, without isolating the reaction from air. This forms the very stable iron (III) oxide as an insoluble material, while dissolving chromates.

The sodium chromate produced in this process is converted into sodium dichromate using sulfuric acid, the by-product being sodium sulfate.

From this point, carbon is used to reduce the dichromate to chromium (III) oxide, which is then reduced to the metal using aluminum as the reducing agent.

The simplified reaction chemistry can be presented as follows:

$$4FeCr_2O_4 + 8Na_2CO_3 + 7O_2 \longrightarrow 8Na_2CrO_4 + 8CO_2 + 2Fe_2O_3$$

$$2Na_2CrO_4 + H_2SO_4 \longrightarrow Na_2Cr_2O_7 + H_2O + Na_2SO_4$$

$$Na_2Cr_2O_7 + 2C \longrightarrow Cr_2O_3 + Na_2CO_3 + CO$$

$$2Al + Cr_2O_3 \longrightarrow Al_2O_3 + 2Cr$$

26.2.3 Major uses

Ferrochromium

Chromium use in steel is vital, as the USGS Mineral Commodity Summaries states, "Chromium has no substitute in stainless steel, the leading end use" [6]. The term 'ferrochrome' indicates an alloy that is at least half chromium. It is produced from chromite ore in an electric arc melting process. But this is itself consumed in stainless steel production, which brings the amount of chromium in the final alloy down to approximately 11 %.

Chrome plating

Chrome plating on metal surfaces has become common, both for its attractiveness, and its long-term resistance to corrosion. The exposed metal parts of automobiles, trucks, and motorcycles are very often chrome coated.

Pigments and dyes

Several yellow and green dyes and pigments used in paints and in glassmaking must incorporate chromium, as a chromium compound, to attain the necessary colors. Some of the more common ones are listed in Chapter 16.

26.3 Mercury

Mercury has been known since ancient times, and since it is the only metal that exists as a liquid at ambient temperature and pressure it has been called quicksilver for much of this time. We have discussed the use of mercury in the production of sodium hydroxide, but have not yet examined the actual production of elemental mercury from its natural sources.

26.3.1 Sources

Cinnabar

Mercury is routinely extracted from the mercuric sulfide ore (HgS) known as cinnabar. Other ores that contain mercury include the rather scarce corderoite ($Hg_3S_2Cl_2$) and livingstonite ($HgSb_4S_8$). While the United States no longer has any operational mines for mercury ores, countries such as Kyrgyzstan and China still have active operations. The McDermitt Mine, in Nevada, USA closed in 1992 [6].

Mercury metal is separated from its ore by heating in the presence of oxygen, and then condensing the resultant vapor. The reaction chemistry can be represented as follows:

$$HgS(s) + O_2(g) \longrightarrow Hg(l) + SO_2(g)$$

Recycling

Recycled mercury is used in many applications within the United States, since there are no mercury mines active in the country. The US also imports the element from Peru, Chile, Germany, and Canada, and currently maintains a government stockpile of 4,436 tons [6].

26.3.2 Volume produced annually

The United States does import elemental mercury. Recent data for imported volumes are shown in Table 26.2.

Table 26.2: Mercury Imports to the United States.

Year	Amt. Imported (mTon)
2007	67
2008	155
2009	206
2010	294
2011	160*

*Recent decline is due to an increase in recycling of mercury from end use materials [6].

26.3.3 Major uses

The Mercury Cell Process, detailed in Chapter 6, remains the largest single use for elemental mercury. Mercury use in specialized, high temperature thermometers, as well as in fluorescent light bulbs, utilizes nowhere near as much of the element, but these two end uses are those with which most consumers are familiar.

26.4 Gold

26.4.1 Introduction

Perhaps no other metal attracts the attention of people like gold. Used since ancient times for adornment and as a store of wealth, gold has been found throughout the world as nuggets that are oftentimes called placers. The nugget simply sits in or on the ground or in a stream bed, and is gathered by a person, then either used directly, or refined for some later use. There have been famous, and not-so-famous gold rushes throughout history. The Line of Tordesillas was drawn in part to prevent wars between Portugal and Spain concerning their claims to the New World, and supposed wealth of it, much of which was believed to be gold. Enough gold was found in the Appalachians twice in the early nineteenth century that the United States government set up branch Mints in Charlotte, North Carolina, and Dahlonega, Georgia exclusively to produce gold coins. The very famous gold rush of 1849 left a permanent mark on the American west (including the name of a football team today – the San Francisco '49'ers), and the Yukon gold rush of the late 1800s sent thousands of prospectors to that cold part of the Canadian north. Even today, as recently as 2012, prospectors are returning to the streams and rivers of California to glean the gold overlooked by the prospectors of the last century [5].

26.4.2 Sources

The USGS Mineral Commodities Summary indicates that gold in the United States is now mined mainly in Alaska, with smaller operations in the continental United States in the west. Further, it indicates that a much smaller amount is recovered as a by-product in copper refining.

The metal is however mined throughout the world. Figure 26.3 shows its production as a percentage, with the whole being 2,700 metric tons.

While this total number is far, far lower than the other metals we have discussed, bear in mind that most metals are priced in pounds, and that gold in the year 2013 has recently come *down* to approximately $1,300 per troy ounce from an earlier high of $1,900 per troy ounce.

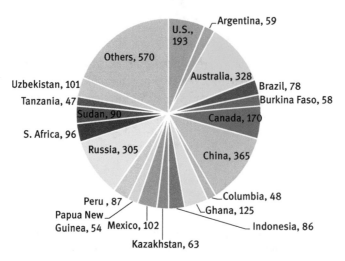

Figure 26.3: Gold production in metric tons.

26.4.3 Extraction chemistry

Aqua regia

Gold is often extracted from ores through dissolving it in aqua regia, a mixture of nitric and hydrochloric acids. The nitric acid is capable of solvating the gold, while the chloride forms the $AuCl_4^-$ anion, which can then be separated and reduced.

Miller process

What is called the Miller Process uses a directed stream of elemental chlorine over impure gold while it is being heated. The impurities form metal chlorides, leaving the gold unreacted. The chloride salts of the impurities are then removed, leaving gold that can be as high as 99.9 % purity.

Anode mud

Industrially, smaller amounts of gold are produced at the anodes of high purity copper electrolytic refining operations, as mentioned in Chapter 25.

Cyanide leaching

Since the late 1800s, extracting gold from ore using cyanide has become a profitable way to recover gold from ores in which the gold is a small portion of the overall mixture. The reaction chemistry can be portrayed fairly simply, as:

$$8NaCN + 4Au + O_2 + 2H_2O \longrightarrow 4Na[Au(CN)_2] + 4NaOH$$

The process will also work with a potassium or calcium cyanide salt. Since the complex ion shown is not a desired end product, the gold must be reduced back to the elemental metal. This unfortunately co-produces large amounts of cyanide-containing waste that must be neutralized, or that becomes an environmental hazard.

Attempts have also been made to extract gold from seawater, although none have yet met with commercial success. The earliest such attempt may still be the most famous: world-famous chemist Fritz Haber led a team immediately after the First World War, in an attempt to recover enough gold from the sea for Germany to pay for the war reparations that had been imposed on the nation by the war's victors.

26.4.4 Major uses

Although many people consider the major use for gold to be a prime material for jewelry or for storing wealth, there are a number of industrial applications as well [7]. Gold is used extensively in computer connections because of its superior conductivity when compared to other metals. When used in small enough amounts per end item, the presence of gold does not bring the price of that product up appreciably.

In the last thirty years, the minting and production of gold coins at standard weights of one troy ounce and fractions thereof has become a means by which individuals can own gold as a store of wealth. The South African Krugerrand was the first "bullion coin," and indeed was the only one for several years. But in the 1980s, several national mints began producing gold bullion coins for the precious metals markets. United States Eagles, Canadian Maple Leaves, Chinese Pandas, Australian Nuggets, and Britannias of Great Britain are all now traded, and several other nations have begun precious metal bullion coin programs, centered on some gold coin.

The allure of gold is such that these bullion coins can not only be purchased from metals dealers and jewelry stores, but in Las Vegas, Nevada there is at least one casino that operates a vending machine that dispenses gold in ingots as small as one gram. Developments like this mean that the price of gold has begun to fluctuate in the past few years. A long-term price comparison is illustrated in Figure 26.4. While the price has dropped in the past year from a very high peak, many speculators believe it will rise to even higher heights in future years.

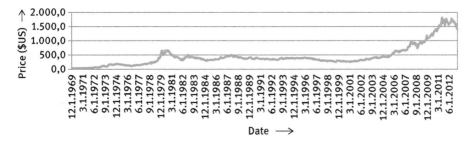

Figure 26.4: Gold Prices Over the Past Decades.

26.5 Silver

26.5.1 Introduction

Much like gold, silver has been a known metal since antiquity, and in some instances has been more highly valued than gold (some ancient civilizations believed gold was an impure form of silver). Silver has been used throughout history for ornamentation, and for items, tools, and utensils that signal the wealth of the owner, but became an important component of photographic prints as that medium developed in the mid-nineteenth century.

Largely because the price of silver has always been lower than gold from the time of the Industrial Revolution, there has always been more industrial applications for silver [4]. National currencies were often fixed in terms of both gold and silver, in some specific ratio of the two metals.

26.5.2 Sources

Silver is mined as a co-product of other metals throughout the world. Figure 26.5 indicates the percentage by country, based on a total of 23,800 metric tons in 2011 [6].

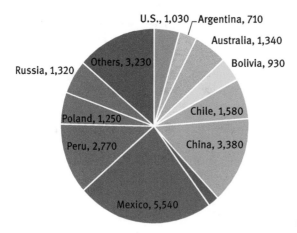

Figure 26.5: Silver Production in metric tons, by country.

26.5.3 Extraction chemistry

Most silver is mined as a co-product of the following: lead-zinc mines, copper mines, as well as gold mines [6]. In the recent past, much of the discovered silver has been in conjunction with gold discoveries. The famous silver mine of Potosi, which produced enough silver that it changed the economy of Europe in the 1500s, is still mined today, although silver is not its main product. The Comstock Lode of Nevada, another famous mine that changed the economy of the United States in the latter part of the nineteenth century, has been mined out. This mine did however attract enough attention that it gave rise to the Washoe Process, a silver recovery technique based on amalgamation.

Washoe process

Batches of 1,200 pounds of crushed ore are placed in copper pans and amalgamated with mercury. Both sodium chloride and copper sulfate are added. The batches are agitated with an iron plate or paddle, which provides small amounts of iron to the mixture. With heating in the form of steam, and agitation, impurities are removed, leaving the silver mercury amalgam, from which the silver is recovered.

26.5.4 Major uses

The uses of silver are numerous [4], but photography, silverware and jewelry, and coins remain major uses at the top of this long list.

Bullion coins

As mentioned when discussing gold, silver is another metal that has been made into bullion coins by several nations in the past thirty years. Silver has throughout history been used for circulating coins, usually as a 90 % silver to 10 % copper alloy. Called "coin silver," this alloy was lower on purpose than sterling silver (92.5 % silver) to prevent counterfeiting of coins by melting silverware. In the 1960s, silver was taken out of the silver coins of most nations, and was replaced with different alloys that had the look and feel of the precious metal. In the 1980s bullion coins were produced from silver for use on world markets. With the exception of the gold Krugerrand, the gold bullion coins mentioned above all have counterparts in silver, usually as one-ounce coins. In the past few years, multi-ounce silver bullion coins have been produced by some world mints, in response to perceived desires by speculators and collectors.

Photography

The use of silver in photographic applications has declined in the past ten years, as the availability of inexpensive digital cameras and direct printing has become more widespread. However, a certain amount of silver is still used for photographs.

Silverware and jewelry

Silver metal continues to be utilized in these two end use materials. Old, damaged, or unwanted silverware and silver jewelry is also recovered at second-hand shops that specialize in precious metals.

26.6 Platinum group metals (PGM)

26.6.1 Introduction

The platinum group metals will be treated as a group. Each of the PGM are rare elements, and the group includes: platinum, palladium, rhodium, ruthenium, iridium, and osmium. While these are scarce metals compared to the large-scale industrial metals, their value and the importance of their niche applications is such that there is a trade organization devoted to their use [1].

26.6.2 Sources

The United States has operating mines in Montana that produce PGMs, but most platinum is imported to the US from: Germany, South Africa, Great Britain, Canada and others [6, 1].

World platinum production in 2011 was 192,000 kilograms, and was produced as is shown, by percentage, in Figure 26.6.

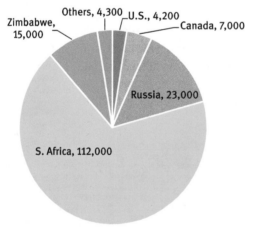

Figure 26.6: Platinum Production in Kilograms.

The Merensky Reef

Although platinum as an element was first discovered in the mines of South America, South African production of platinum and other PGM now dominates the world market. South African production has itself been dominated since the 1950s by the production at the Merensky Reef, a localized area within the Bushveld Igneous Complex [1, 3].

26.6.3 Extraction chemistry

Platinum is often extracted from other metals by dissolving everything that is not platinum in acid mixtures. The platinum can then be removed from the bottom of the mixture and separated.

As mentioned in the discussion of gold extraction, platinum can also be recovered through the electrolytic refining of copper. It is extracted from the anode muds that form when copper is being purified.

26.6.4 Major uses

Platinum finds its greatest commercial and industrial use in catalytic converters in automobiles [3]. Platinum is also used for jewelry and bullion coins as another way to store wealth. The bullion coin programs mentioned for gold and silver extend to platinum as well, at least in the United States, Canada, China, and Australia. In the last decade, other countries have ventured into producing platinum bullion coins as well.

26.7 Uranium

Uranium is well known for its two major uses: the product of fuel rods for nuclear fission power plants, and the production of fissile material for atomic weaponry.

26.7.1 Introduction and history

The discovery and history of uranium is closely tied to several of the famous chemists and physicists at the turn of the 20th century. Henri Becquerel was the first to discover the radioactive nature of the element, by inadvertently exposing undeveloped photographic plates to a uranium-bearing salt. Marie Curie extracted a few grams of the element radium from several tons of uranium-containing pitchblende ore, and was the first to use the term "radioactivity."

26.7.2 Sources

Uranium ores are found at the following places throughout the world, as seen in Table 26.3. A great deal of uranium deposits are in pitchblende ores, although other ores contain the element as well.

While some ores are as low in concentration as 0.1%, the ores from the Canadian Rockies can be as high as 20% uranium oxide, which means much less refining is re-

Table 26.3: Uranium Ore Locations.

Country	Location	Characteristics and comments
Kazakhstan		Sandstone deposits
Canada	McArthur River	High grade deposits
Australia	Northern Territory	High grade deposits
Namibia	Langer Heinrich Uranium mine	Extracted as raw U_3O_8
Niger	Agadez, northern Niger	Mining through French-led conglomerate, extracted as raw U_3O_8
Russia	Kurgan Region	Sandstone deposits
USA	Wyoming, New Mexico	Sandstone deposits

quired to isolate the metal, or usable UO_2. The normal refining technique involves the following steps:
1. Crushing the ore to a powder.
2. Dissolving the powder in acid or base.
3. Precipitation and extraction (this step may need to be repeated several times).
4. Ion exchange (which again may need several cycles).

The product, generally U_3O_8, but a mixture of uranium oxides and other salts – called yellowcake – is then smelted to obtain UO_2. This can be used in uranium fuel rods for what are termed un-enriched reactors.

Isotopic enrichment to obtain higher concentrations of U-235 routinely starts with U_3O_8, and reacts it with fluorine gas to create $UF_6(g)$. Gas diffusion was the first method of separation, and was used by the United States in the Manhattan project, during the Second World War. This technique has in large part been replaced by high speed centrifugation in recent years. Because the separation is dependent on the atomic mass of the products, and the two materials to be separated are UF_6 containing U-235 and U-238, the separation process is very energy intensive, since there is such a slight atomic mass difference. Obtaining approximately 20 % U-235 in a fraction is sufficient for nuclear power use, although for compact reactors (onboard large, military ships, for instance) higher enrichment is required. Even higher enrichment is required for what is known as weapons grade uranium.

26.7.3 Major uses

To a lesser extent than its uses in power plants and weaponry, uranium has been used as a glass additive to produce red and red-orange colors.

26.7.4 Recycling

While national governments do not always track the recovery of these metals, scrap yards are usually willing to do business in all of them (with the exception of uranium, which is often tracked and controlled nationally). Gold and silver recycling are cottage businesses throughout much of the developed world, with shop owners then selling to businesses and industries which possess the equipment and furnaces necessary to re-melt and possibly refine the metals. There are also several companies that work in mercury recovery. They then re-sell the metal to industrial and niche users.

Bibliography

[1] International Platinum Group Metals Association. Website. (Accessed 10 December 2022, at www.ipanews.com/en/).
[2] International Titanium Association. Website. (Accessed 10 December 2022, at www.titanium.org/).
[3] Platinum Today. Website. (Accessed 10 December 2022, at matthey.com/products-and-markets/pgms-and-circularity/pgm-management).
[4] The Silver Institute. Website. (Accessed 10 December 2022, at www.silverinstitute.org/site/).
[5] Tucker, A. Going for the Gold, Smithsonian, July–August, 2012, 34–36.
[6] USGS Mineral Commodity Summaries 2022, US Department of the Interior, U. S. Geological Survey, https://doi.org/10.3133/mcs2022.
[7] World Gold Council. Website. (Accessed 10 December 2022, at www.gold.org/).

27 Materials

27.1 Introduction, silicates

The broad term "silicates" encompasses many materials with many uses, from gem-stones to enamel. We will not be able to treat them all here, but will discuss enamel and ceramics, both of which incorporate silicates in a variety of ways. Earlier in this book, we have listed several silicates in examples of various inorganic pigments.

27.2 Enamel

27.2.1 Introduction

What is often called vitreous enamel or porcelain enamel is a glass coating, often applied on a metal surface, that begins as powdered glass, and that is transformed to a solid, smooth coating by firing the surface at temperatures in the range of 750–850 °C.

27.2.2 Production

There is a wide enough manufacturing base that a Porcelain Manufacturing Institute exists [8], to promote communication and spread developments throughout the industry.

On its website, the PEI claims its goals are to: "Showcase and promote innovations in materials and processing to improve the overall efficiency of enameling operations. Promote the product generally and encourage its use in all possible applications. Advance and protect the legitimate interests of the industry and its individual members." [8] As well, there is a Deutsche Email Verband e. V., which is the oldest professional union of in-dustries dealing with the enameling of products [2]. In addition, there is an International Enamellers Institute, [4] another trade organization that deals with industrial aspects of vitreous enameling.

Enamel surfaces for end use items are produced by applying a ground, powdered glass layer upon a prepared surface, often a metal surface. Metal surfaces are almost always rigidly degreased, and some still require a very mild etching to produce a surface to with the glass can bond while molten.

Often, a second coat is applied to the first enamel coat. The second coat is usually clear or white, and protects the first coat, often keeping it looking attractive over long periods of time.

https://doi.org/10.1515/9783110671094-027

27.2.3 Uses of enamel

Many people tend to think of enamel as an attractive finish for art objects, and indeed there is some use of enamel for artistic purposes. But the production of numerous end use items, including bathtubs and showers, sinks, and toilets all can include enamel. The Porcelain Manufacturing Institute lists in some detail other applications beyond these common ones [8], including food processing equipment and package chutes.

27.3 Ceramics

27.3.1 Introduction and classifications

The word "ceramics" quickly brings to mind various types of pottery, but this is only one small application for a diverse array of materials. The field is large enough that an American Ceramics Society [1] brings users and manufacturers together to examine the latest industry developments, among other things. Ceramics are oftentimes silicate in some component, and are often crystalline or semi-crystalline. Ceramics can be amorphous in their microscopic structure, but amorphous materials are more properly referred to as glasses, and thus can be considered enamels (at least for the course of our discussion in this chapter).

In general, ceramics are divided into the following: structural and building materials, consumer end use products such as cookware and pottery, and linings for high temperature operations. Of course, there are several other specialty applications.

27.3.2 Production

There is no unified theory of ceramic composition. Rather, both decorative and structural uses have been determined through trial-and-error of many different compositions. What most ceramic production has in common, despite differences in chemical make-up, is the shaping of the object, often with some amount of solvent involved (not always water), then firing of the object in some form of oven or kiln, followed by slow cooling.

Beyond the formation of ceramic objects, there are numerous end use products that must then be coated, or glazed, and subjected to a second firing and cooling sequence. While many types of pottery are made this way, there are many objects which are produced that are not solely decorative, but that must be produced with multiple firings.

27.3.3 Major uses

Ceramics are used in a very wide variety of end us products. Domestic products are certainly well known, but there are other as well that are usually not considered very often. These include:
– Roofing tile
– Floor tile
– Wall facing
– Brick work
– Tableware and cookware
– Kiln and oven linings
– Ballistic body armor inserts
– Armored vehicle plating and reinforcement

One famous use of ceramic plating is the skin of the United States space shuttles, but this is certainly not a large volume operation when viewed in comparison with many of the others we have discussed. Precisely because there are so many uses of ceramics, it is difficult to put them into categories, without realizing that many of them could be placed in more than one.

27.4 Metal foams

Very recently, a great deal of research has gone into the production of what are called metal foams [6]. A metal foam is a material in which the metal is positioned much like the solid material of a sponge, and again like a sponge, the remaining space is occupied by air. Unlike a traditional sponge however, the metal framework is generally rigid. Thus, materials have been made in which the density of the material is very low because the framework metal is interlaced with a significant amount of air, sometimes greater than 75 %. In general, metal foams are divided into what are called closed cell and open cell foams. Most are made using aluminum, since it is already a lightweight metal.

27.4.1 Production

Injection of a gas or of some other foaming agent into a metal while it is molten produces a metal foam upon cooling. Care must be taken during production to ensure the bubbles remain in the material long enough for the foam to form. Injecting a gas into molten metal is usually a technique used when the metal exists as a viscous alloy in the melt. As an alternative method of production, the bulk material can be melted after some liquid or solid has been premixed, this mix then being able to release gas into the molten material.

27.4.2 Uses

Metal foams find use as heat exchangers and shock absorbent materials. Despite an established history for these materials (the first patents for them were issued in the late 1940s), they remain expensive to produce, and thus tend to occupy only certain niche applications. For example, the aerospace industry requires lightweight materials in a variety of applications. Metal foams work well in some of these [3].

27.5 Carbon materials

27.5.1 Introduction

The discovery of buckminsterfullerene, and the rapid pace of research related to what has become known as the fullerenes, is one of the success stories of the past twenty years. Seldom is a new allotrope of an element discovered and verified in so short a time. To have moved from the discovery stage to the first small companies producing fullerenes for any use in this short a span of years is quite amazing. Figure 27.1 shows the structure of the simplest fullerene.

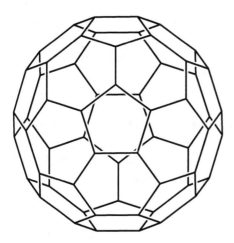

Figure 27.1: Structure of C_{60}, the Simplest Fullerene.

27.5.2 Bucky ball fullerenes

The first and smallest of the fullerenes, C_{60}, has generated a great deal of interest and research, especially after it was proven that atoms can be enclosed in the structure of the molecule, as a sort of clathrate. This form of material, generally referred to as an endohedral fullerene, uses the "@" notation to indicate the atom or particle within the cage. For example, the first such compound was a lanthanum within the cage, and is written: La@C_{60}.

Fullerenes have been produced by electrolysis of a graphite rod in a vacuum. The production of endohedral fullerenes involves having the atom or particle slated for inclusion to be present in the otherwise evacuated chamber when the electrolysis of the carbon is initiated. The end result is the production of both fullerene and endohedral fullerene, which must be separated. To date, none of these materials has been used in a large scale industrial process.

27.5.3 Tubular fullerenes

The development of methods by which larger fullerenes, something called buckytubes and sometimes called graphene tubes, can be made has occurred very rapidly, and holds commercial potential. The materials may prove to be useful catalysts or catalyst supports, or may prove to be equally useful in medicine, in the field of time release drugs.

27.5.4 Major uses

Currently, there are no uses of fullerenes, buckyballs or buckytubes that have become large corporate endeavors. Producers of various fullerenes exist, and advertise these materials for sale and use in other applications [7, 5].

While fullerene production has not yet progressed to large scale operations, there are however numerous websites promoting specific uses of fullerenes, including those claiming that buckminsterfullerene in olive oil has anti-aging properties.

27.6 Pollution, recycling, and by-product uses

The production of many of the materials discussed here is energy intensive, since high temperatures are required to melt the material during manufacture. Beyond that, these processes can be clean, and relatively pollutant free.

There are no recycling programs for these materials, as many of the end use items are made to last for at least one human lifetime, if not more. While there is recycling for glass containers, this is reserved to bottles and other glass that is mass produced to certain specifications. Most of the materials in this chapter are not produced on a large enough scale that any recycling would be profitable.

Bibliography

[1] American Ceramics Society. Website. (Accessed 10 December 2022, at ceramics.org/).

[2] Deutsche Email Verband, e.V. Website. (Accessed 10 December 2022, at www.emailverband.de/General/de/DEV/DEV_Index.php).

[3] Dukhan, Nihad. Metal Foams: Fundamentals and Applications. DEStech Publications, Lancaster, PA, USA, 2012. 978-1-60595-014-3.

[4] IEI, International Enamellers Institute. Website. (Accessed 10 December 2022 at ieiworlddotorg.wordpress.com).

[5] Mer Corporation. Website. (Accessed 10 December 2022, at linkedin.com/company/mer-corporation/about).

[6] Metal Foams. Website. (Accessed 10 December 2022, at www.metalfoam.net/).

[7] Nano-c Nanostructured Carbon. Website. (Accessed 10 December 2022, at www.nano-c.com/).

[8] The Porcelain Enamel Institute, Inc. Website. (Accessed 10 December 2022, at www.porcelainenamel.com).

Index

https://doi.org/10.1515/9783110671094-028

Printed in the USA
CPSIA information can be obtained
at www.ICGtesting.com
LVHW081353010624
781922LV00007B/640

9 783110 671063